Nuclear Radiation Detection
Materials—2011

MATERIALS RESEARCH SOCIETY
SYMPOSIUM PROCEEDINGS VOLUME 1341

Nuclear Radiation Detection Materials–2011

Symposium held April 25–29, 2011, San Francisco, California, U.S.A.

EDITORS

Michael Fiederle

Albert-Ludwigs-Universitaet Freiburg
Freiburg, Germany

Arnold Burger

Fisk University
Nashville, Tennessee, U.S.A

Larry Franks

Special Technologies Laboratory
Santa Barbara, California, U.S.A.

Kelvin Lynn

Washington State University
Pullman, Washington, U.S.A.

Dale L. Perry

Lawrence Berkeley National Laboratory
Berkeley, California, U.S.A.

Kazuhito Yasuda

Nagoya Institute of Technology
Nagoya, Japan

Materials Research Society
Warrendale, Pennsylvania

CAMBRIDGE
UNIVERSITY PRESS

Shaftesbury Road, Cambridge CB2 8EA, United Kingdom

One Liberty Plaza, 20th Floor, New York, NY 10006, USA

477 Williamstown Road, Port Melbourne, VIC 3207, Australia

314–321, 3rd Floor, Plot 3, Splendor Forum, Jasola District Centre, New Delhi – 110025, India

103 Penang Road, #05–06/07, Visioncrest Commercial, Singapore 238467

Cambridge University Press is part of Cambridge University Press & Assessment, a department of the University of Cambridge.

We share the University's mission to contribute to society through the pursuit of education, learning and research at the highest international levels of excellence.

www.cambridge.org
Information on this title: www.cambridge.org/9781605113180

Materials Research Society
506 Keystone Drive, Warrendale, PA 15086, USA
http://www.mrs.org

First published 2012

CODEN: MRSPDH

A catalogue record for this publication is available from the British Library

ISBN   978-1-605-11318-0   Hardback

CONTENTS

*Invited Paper

OTHER MATERIALS

*Invited Paper

PREFACE

Symposium U, "Nuclear Radiation Detection Materials," held April 26–28 at the 2011 MRS Spring Meeting in San Francisco, California, provides the latest results and discussion of nuclear radiation detection materials. The symposium gave an overview of the crystal growth of the radiation detector materials, the characterization and the technology issues.

There is a strong need for new materials and methods for a variety of radiation detection applications. The types of detector materials include semiconductors and scintillators, which are represented by a variety of new scintillator materials, novel semiconductors, and traditional detection materials. This symposium was the continuation of the symposium from 2009 and it shows a rapidly growing field with several important improvements for the development of future radiation detectors.

Michael Fiederle
Dale L. Perry
Arnold Burger
Larry Franks
Kazuhito Yasuda
Kelvin Lynn

November 2011

MATERIALS RESEARCH SOCIETY SYMPOSIUM PROCEEDINGS

MATERIALS RESEARCH SOCIETY SYMPOSIUM PROCEEDINGS

Prior Materials Research Society Symposium Proceedings available by contacting Materials Research Society

Scintillators

Mater. Res. Soc. Symp. Proc. Vol. 1341 © 2011 Materials Research Society
DOI: 10.1557/opl.2011.1205

Transparent Lu_2O_3:Eu Ceramics

Zachary M. Seeley, Joshua D. Kuntz, Nerine J. Cherepy, and Stephen A. Payne

Chemical Sciences Division, Lawrence Livermore National Laboratory, Livermore, CA 94550-9698, U.S.A.

ABSTRACT

We are developing highly transparent ceramic oxide scintillators for high energy (MeV) radiography screens. Lutetium oxide doped with europium (Lu_2O_3:Eu) is the material of choice due to its high light yield and stopping power. As an alternative to hot-pressing, we are utilizing vacuum sintering followed by hot isostatic pressing (HIP). Nano-scale starting powder was uniaxially pressed into compacts and then sintered under high vacuum, followed by HIP'ing. Vacuum sintering temperature proved to be a critical parameter in order to obtain highly transparent Lu_2O_3:Eu. Under-sintering resulted in open porosity disabling the driving force for densification during HIP'ing, while over-sintering lead to trapped pores in the Lu_2O_3:Eu grain interiors. Optimal vacuum sintering conditions allowed the pores to remain mobile during the subsequent HIP'ing step which provided enough pressure to close the pores completely resulting in fully-dense highly transparent ceramics. Currently, we have produced 3 mm thick by 4.5 cm diameter ceramics with excellent transparency, and anticipate scaling to larger sizes while maintaining comparable optical properties.

INTRODUCTION

Transparent polycrystalline ceramics have gained significant interest for applications in laser host materials, high index lenses, transparent armor, radiation detectors, and radiography screens [1-5]. They can be formed in a wider range of compositions, sizes, shapes, and at lower cost than single crystals. Transparent ceramics are however limited to cubic crystal structure materials, and optimized processing is necessary to achieve full density and transparency.

Lutetium oxide doped with europium (Lu_2O_3:Eu) has become a material of interest as a scintillating radiography screen due to its high density and x-ray stopping power, efficient conversion to visible light, and visible emission at ~600 nm coupling well with silicon CCDs [6,7]. Due to the high melting point of Lu_2O_3 (~2400°C), sintering to full density is challenging. Conventionally, hot-pressing overcomes this obstacle by applying pressure and temperature simultaneously [8]. However, along with this technique come a highly reducing environment and potential for contamination from the graphite tooling, requiring a post-treatment anneal which often degrades the transparency.

In this work, we have employed the sinter-HIP method to obtain highly transparent Lu_2O_3. Under this process, the ceramic is vacuum sintered to closed porosity and then subsequently HIP'ed under inert argon gas pressure to full density. This method allows consolidation in a less reducing environment and lower potential for contamination.

EXPERIMENT

Lu$_2$O$_3$ doped with 5at.% Eu (Lu$_{1.9}$Eu$_{0.1}$O$_3$) nanopowder was synthesized via the flame spray pyrolysis (FSP) method by Nanocerox[TM] (Ann Arbor, MI). As received powder had a specific surface area of 22 m^2/g and was crystallized in the cubic bixbyite structure.

Nanopowder was suspended in an aqueous solution of polyethylene glycol (PEG) and ammonium polymethacrylate using an ultrasonic probe (Cole Parmer, Vernon Hills, Il) and a high shear mixer (Thinky, Japan). This suspension was spray dried (Buchi, New Castle, DE) at 210°C into flowing nitrogen to protect the organics. The dried powder was then sieved (<50μm) to create uniform agglomerates of nanoparticles with an even distribution of organic additives. Formulated nanoparticles were then uniaxially pressed at 50 MPa to form green compacts approximately 35% dense, followed by calcination in air to remove the organics. Calcined compacts were then loaded into a tungsten element vacuum furnace (Thermal Technologies, Santa Rosa, CA) and sintered under a vacuum of <2×10^{-6} Torr at temperatures ranging between 1575 and 1850°C. The sintered structures were then hot isostatically pressed (HIP'ed) under 200 MPa of inert argon gas pressure at 1850°C for 4 h in a tungsten element HIP (American Isostatic Presses, Columbus, OH). Ceramic surfaces were then ground flat and parallel and given an inspection polish to qualify transparencies. Optical microscopy was used to characterize transparency on a micrometer scale.

RESULTS AND DISCUSSION

The sinter-HIP methodology for densifying powder compacts is a two step process by which the compacts are first vacuum sintered to closed porosity followed by HIP'ing under inert argon gas pressure at high temperatures [9]. During the vacuum sinter step the initial and intermediate stages of sintering are occurring. First necks form between particles. During the intermediate sintering stage, the necks are growing, forming grain boundaries, and pores begin to shrink causing sample densification. With sufficient densification, the pores between the grains are no longer interconnected at which point the compact reaches closed porosity. This step is performed in a high vacuum so that when the pores close to the external surface only vacuum remains trapped in the pores. At this point, external pressure can be applied in the form of an inert gas (argon) at high temperature without the gas infiltrating the porosity of the compact. The pressure provides a secondary driving force for material to diffuse into the vacuum filled pores. In the final stage of sintering the porosity is eliminated and grains begin to grow forming the fully dense ceramic [10].

In the sinter-HIP process, the vacuum sintering temperature is critical to optimize in order to form transparent ceramics. Figure 1 shows photographs of vacuum sintered Lu$_2$O$_3$:Eu before and after the HIP'ing step to show the importance of optimizing the vacuum sintering temperature. The sample sintered at 1575°C clearly did not have closed porosity. During the HIP'ing step the argon infiltrated the pores of this sample resulting in very little densification. Samples vacuum sintered between 1600 and 1650°C show high optical transparency after the HIP'ing step, but as the vacuum sintering temperature increased the transparency degrades due to over-sintering. Vacuum sintering at 1600°C was enough to reach closed porosity for the given green density yet it was low enough not to initiate the final stage of sintering. As the vacuum sintering temperature increased up to 1850°C the final stage of sintering begins to occur

4

accompanied by grain growth. In vacuum sintered samples with larger grain size the grain boundary area decreases allowing fewer pathways for atomic diffusion and pore removal during the HIP'ing step.

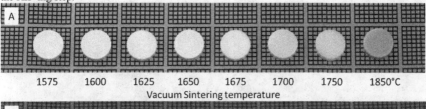

| 1575 | 1600 | 1625 | 1650 | 1675 | 1700 | 1750 | 1850°C |

Vacuum Sintering temperature

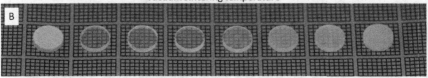

Figure 1. (A) Photograph of compacts as a function of the vacuum sintering temperature between 1575 and 1850°C, and (B) the same compacts after HIP'ing at 1850°C. Samples display a red-orange color luminescence under UV excitation.

To better understand the loss of transparency with increased sintering temperature, samples were characterized optically on the micron scale. Figure 2 shows optical micrographs of HIP'ed samples that were vacuum sintered between 1625 and 1700°C. The focal plane of the optical microscope was focused into the interior of the samples to view the bulk scattering defects, therefore only some of the defects are in focus and others are above and below the focal plane. In the sample vacuum sintered at 1625°C, very few pores are visible and only a slight reflection of light at the grain boundaries can be seen. As the vacuum sintering temperatures increases the number of visible defects increases correspondingly. When the vacuum sintering temperature reaches 1675°C small agglomerations of pores are noticed and by 1700°C the center of each grain is filled with a large pore agglomeration. However, no pores are visible near the grain boundaries. This result indicates that vacuum sintering at higher temperatures resulted in pores becoming entrapped in the interior of each grain. Then, during the subsequent HIP step, residual porosity at the grain boundaries was removed but grain-entrapped pores were not. Vacuum sintering at lower temperatures (i.e. 1625°C) resulted in the ideal microstructure of small grains with pores located on the grain boundaries and easily removed during the HIP'ing step.

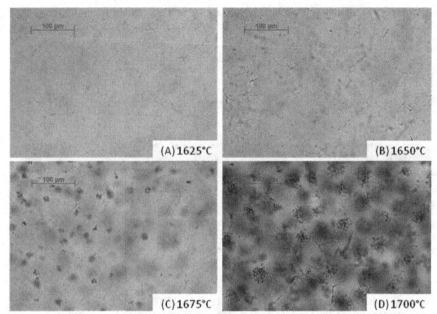

Figure 2. Optical micrographs focused into the interior of the transparent HIP'ed Lu$_2$O$_3$:Eu as a function of the vacuum sintering temperature (A) 1625, (B) 1650, (C) 1675, and (D) 1700°C.

The scintillator screen size determines the maximum object size and magnification that can be imaged by X-ray CT. Therefore, we are utilizing this sinter-HIP processing method to scale up to a size useful in industrial radiography screens. Figure 3 shows a 44g part that is ~4.5 cm in diameter and has equivalent transparency to the smaller test coupons. Larger diameter ceramics are in progress.

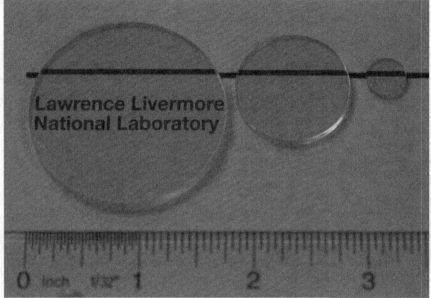

Figure 3. Photograph of Lu_2O_3-based transparent ceramics. The fabrication method described herein allows increased part size while maintaining the high degree of transparency.

CONCLUSIONS

Transparent Lu_2O_3:Eu ceramics were fabricated via the vacuum sinter HIP method. Vacuum sintering temperature proved to be a critical parameter in order to achieve high transparency. Under sintering resulted in open porosity rendering the HIP'ing step ineffective, while over sintering resulted in grain-entrapped porosity with low mobility during the HIP step. The best processing conditions are being used to increase maximum part size for large area scintillator screens.

ACKNOWLEDGMENTS

Thanks to Todd Stefanik of Nanocerox Inc., Jeff Roberts for flame spray synthesis, and Scott Fisher for mechanical fabrication. This work performed under the auspices of the U.S. Department of Energy by Lawrence Livermore National Laboratory under Contract DE-AC52-07NA27344 and funded by the US DOE, Office of NNSA, Enhanced Surveillance Subprogram (Patrick Allen). LLNL-PROC-480414

REFERENCES

1. A. Ikesue, Y. Aung, T. Yoda, S. Nakayama, T. Kamimura, Optical Materials 29, 1289 (2007).
2. U. Peuchert, Y. Okano, Y. Menke, S. Reichel, A. Ikesue, J. European Ceram. Soc. 29, 283 (2009).
3. R. Klement, S. Rolc, R. Mikulikova, J. Krestan, J. European Ceram. Soc. 28, 1091 (2008).
4. N. Cherepy, S. Payne, S. Asztalos, G. Hull, J. Kuntz, T. Niedermayr, S. Pimputkar, J. Roberts, R. Sanner, T. Tillotson, E. Loef, C. Wilson, K. Shah, U. Roy, R. Hawrami, A. Burger, L. Boatner, W. Choong, W. Moses, IEEE Tran. Nuc. Sci., 56, 873 (2009).
5. T. T. Farman, R. H. Vandre, J. C. Pajak, S. R. Miller, A. Lempicki, A. G. Farman, Oral Surg. Oral Med. Oral Pathol. Oral Radiol. Endod. 101, 219 (2006).
6. A. Lempicki, C. Brecher, P. Szupryczynski, H. Lingertat, V. Nagarkar, S. Tipnis, S. Miller, Nuc. Inst. Meth. Phys. Res. A 488, 579 (2002).
7. Y. Shi, Q. W. Chen, J. L. Shi, Optical Materials 31, 729 (2009).
8. D. J. Wisniewski, L. A. Boatner, J. S. Neal, G. E. Jellison, J. O. Ramey, A. North, M. Wisniewska, A. E. Payzant, J. Y. Howe, A. Lempicki, C. Brecher, J. Glodo, IEEE Trans. Nuc. Sci. 55, 1501 (2008).
9. J. Echeberria, J. Tarazona, J. He, T. Butler, F. Castro, J. European Ceram. Soc. 22, 1801 (2002).
10. R. M. German, *Powder Metallurgy Science*, 2nd ed., Metal Powder Industries Federation, Ney Jersey, (1984) p. 261-264.

Mater. Res. Soc. Symp. Proc. Vol. 1341 © 2011 Materials Research Society
DOI: 10.1557/opl.2011.1100

Nuclear Radiation Detection Scintillators based on ZnSe(Te) crystals.

Volodymyr D. Ryzhikov

Institute for Scintillation Materials of STC "Institute for Single Crystals" NAS of Ukraine,

60 Lenin Ave., Kharkov, 61001, Ukraine

ABSTRACT

We describe development of semiconductor scintillators (SCS) on the basis of $A^{II}B^{VI}$ compounds has bridged the gap in a series of "scintillator-photodiode" detectors used in modern multi-channel low-energy devices for visualization of hidden images (tomographs, introscopes). In accordance with the requirements of eventual applications, such SCS materials as ZnSe(Te) show the best matching of intrinsic radiation spectra to photosensitivity spectra of silicon photodiodes (PD) among the materials of similar kind. They are characterized by high radiation and thermal stability of their output parameters, as well as by high conversion efficiency. In this work, a thermodynamic model is described for interaction of isovalent dopants (IVD) with intrinsic point defects of $A^{II}B^{VI}$ semiconductor structures at different ratios of their charges, a decisive role of IVD is shown in formation of the luminescence centers, kinetics of solid-phase reactions and the role of a gas medium are considered under real preparation conditions of ZnSe(Te) scintillation crystals, and luminescence mechanisms in IVD-doped SCS are discussed.

INTRODUCTION

The use of $A^{II}B^{VI}$ compounds, namely, CdS(Te), as highly efficient scintillators was first proposed by J.Thomas e.a. [1,2]. They also assumed that the scintillation mechanism could be determined by radiative recombination centers on isoelectronic traps. Otherwise, in [3] it was shown that luminescence related to isoelectronic traps (IET) is of exciton character, and the corresponding emission spectra, e.g., in ZnTe(O) and $A^{III}B^{V}$ compounds are narrow discrete lines, as distinct from broad diffuse bands characteristic for the centers involving intrinsic defects. The luminescence maximums and the character of the bands are similar to centers involving defects introduced by other means, such as radiation damage [4,5] and thermal treatment [6].

Thus, an isovalent dopant (IVD) atom, differing from the substituted atom by its ion radius and electronegativity, stimulates formation of defects in the neighboring sublattice. In particular, J.Watkins [7] showed by EPR methods that introduction of tellurium into zinc selenide leads to formation of vacancies $Zn(V_{Zn})$; the displaced Zn moves to the interstitial position, and the V_{Zn} - Zn_i complex, stable up to 400 K, is formed in the vicinity of tellurium atom.

Alongside CdS(Te), other scintillators with IVD are of great interest. We were the first to obtain scintillator ZnSe(Te) [8,9]. As distinct from CdS(Te), its luminescence efficiency is comparable to or higher than in traditional materials like CsI(Tl). At present, ZnSe(Te) scintillators are widely used in inspection equipment.

In a series of studies ([10-12] and others] we have considered in detail thermodynamics of defect formation, preparation methods of ZnSe(Te) and its main characteristics. These results are briefly summarized in the present work.

EXPERIMENT

Preparation technology of scintillation crystals ZnSe(Te) is based on the known method of crystal growth from the melt [12] using the Bridgman technique in vertical compression, furnaces under inert gas (argon) pressure up to $5 \cdot 10^6$ Pa. Growth rate is $2 - 5$ mm/hour, with temperature in the melt zone – up to 1850 K; crystals are grown in graphite crucibles. After growth, crystals are annealed in a vapour of Zn for enhancement and spectral stabilization of luminescence in the $610 - 640$ nm region.

To determine optimum technological parameters of the preparation process of ZnSe(Te), we have studied in detail physical chemistry, thermodynamics and kinetics of interaction of the components in the system ZnSe – ZnTe, by accounting for the composition of the gaseous medium and construction materials of the growth equipment [12-14]. The technology developed by us allows one to obtain reproducibly two types of ZnSe(Te) crystals (mass up to 600 g, diameter up to 40 mm), which are called conventionally "fast" and "slow" scintillators. The main characteristics of there crystals are presented in Tables I.

Table I. Main parameters of scintillation crystals ZnSe(Te).

Parameter	Value
Melting point, K	1773-1793 (depending on [Te])
Density ρ, g/cm^3	5.42
Effective atomic number, Z_{eff}	33
Emission maximum λ_m at 300 K, nm: "fast" scintillator "slow" scintillator	610 640
Refractive index for $\lambda_m = 610$-640 nm	$2.58 - 2.61$
Attenuation coefficient at λ_m, cm^{-1}	$0.05 - 0.15$
Decay time, τ, μs: "fast" scintillator "slow" scintillator	$1 - 3$ $30 - 70$
Afterglow, %	< 0.05 after 3 ms
Light yield, photons/MeV γ	$8 \cdot 10^4$
Light output in relation to CsI(Tl), % for X-rays with E<100 keV (CsI(Tl)=100): at 4 mm thickness at 2 mm thickness	up to 100 up to 170
Matching coefficient between scintsllator and photodiode	up to 0.9

When crystals of $A^{II}B^{VI}$ compounds are grown from the melt under pressure, deviations from stoichiometry in the grown crystals can be larger than 1%. As growth is carried out in semipermeable graphite crucibles, the stoichiometry of the solid crystalline phase is affected by two independent processes — diffusion of the initial charge components through crucible walls and their evaporation.

These processes lead to formation of an ensemble of intrinsic point defects (IPD) of the crystal structure. These defects can be interstitial atoms of metal M_I or chalcogene M_X, vacancies of both types V_M, V_X, anti-structural defects M_X, X_M, complexes of defects involving impurity atoms, with each defect able to exist in several charge states. It is considered that perturbing effects of IPD are localized and extend to several interatomic distances. Formation of vacancies is energetically more favorable in the process of crystallization, but when diffusion processes are predominant, in the case when the melt or crystal are in the atmosphere with excess of one of the components, interstitial IPD are mostly formed, Figure 1.

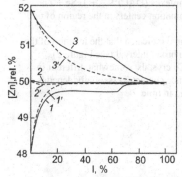

Figure 1. Variation of composition along crystal length in crystals grown by zone melting (1, 2, 3) and directional crystallization (1', 2', 3') at 1800 K and p = 2 MPa, $\Delta z = 1$ mm, v = 19.5 μm/s; $x_0 = 0.48$ (1, 1'), 0.50 (2, 2'), 0.52 (3,3'). [15]

DISCUSSION

Though IVD have the same valence as the substituted lattice atom, such important parameters as ionization energy, electronegativity, ionicity of bonds with atoms of the neighboring sublattice, ion and covalent radii can be substantially different. Bond energy of charge carriers with IVD is by an order of magnitude lower than with donor (D) or acceptor (A) dopants; at the same time, charge localization on IVD is much higher. This combination results in appearance of isolated local states inside the band gap (upon introduction of IVD of the 1st type) with energy levels serving as centers of quenching or emission, or changes in the zone spectrum of allowed bands (IVD of the 2nd type). A criterion for formation of local levels in the band gap is the difference between potential energy values of the IVD atom and the substituted atom [12,16,17]:

$$\Delta U = U_{IVD} - U_{atom} = \Delta E_g - \frac{\partial E_g \big|_{x=1}}{\partial \ln V} \frac{\Delta V}{V}, \tag{1}$$

where $\Delta E_g = E_g|_{x=0} - E_g|_{x=1}$ for compound $A_{1-x}A'_xB(AB_{1-x}B'_x)$; $\Delta V/V$ — relative change in the elementary cell volume from $x = 0$ to $x = 1$.

In the atom pseudopotential model approximation, one should expect local levels in the band gap for crystals CdS(Te), CdSe(Te), ZnSe(Te), ZnTe(Cd) and some others according to the condition

$$(2\Delta U / E_V) \gg 1 \tag{2}$$

where E_V is the band gap width.

It was assumed earlier [12-14] that donor-acceptor complexes (DAC) V$_{Zn}$+ TeSe + Zn$_i$ based on the close pair V$_{Zn}$ – Zn$_i$ could be emission recombination centers in the region of 640 nm at 300 K.

Our investigations of ZnSe doping with various IVD have indicated that the location of the radiation spectrum maximum and its intensity varies within a broad interval (see Figure 2). Finally, preference is given to the ZnSe (Te) and ZnSe (Te, O) crystals, somewhat different in the spectrum and kinetics of the highlighting. The luminescence character essentially depends on the pumping density, the excitation type and it is transforming in time.

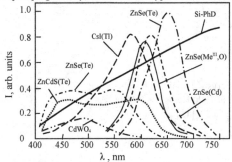

Figure 2. Spectral features of radiation of A^2B^6 scintillators.

It is usually high stable level lightout put from temperature in Te doped crystals ZnSe. At range T=4.2 – 350 K it parameter deviated only at 20 – 30 % .

At difference at crystals doped by O at room temperature luminescence with maximum 600 – 605 nm, and it's intensity droped at 2 – 3 time from 80 to 300 K. Maximum at 80K shifted to 630 nm.

Time decay significantly different so as Figure 3 [18,19].

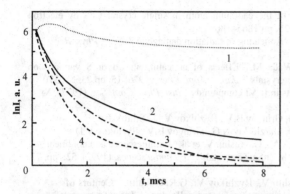

Figure 3. Time decay for scintillator ZnSe with different doped:
1 – ZnSe(Te); 2 – ZnSe(Cd); 3 – ZnSe(O,Al); 4 – ZnSe(O) (T = 300 K).

CONCLUSIONS

It was created new scintillators ZnSe(Te) in the Institute for Single Crystals. About twenty years ago this material wide used in first generations Russian inspection X-ray scanner (Moscow) and medical X-ray CT, St-Petersburg, NIIEFA. The preference of this one is enough high light output, good agreement emission with spectrum sensitivity of Si-photodiode, low afterglow, high radiation stability, no hygroscope.

The disadvantage of this scintillator – badly transparence to own emission (10^{-1} cm^{-1}), long time decay (30 – 50 μs), low density and Z_{eff}. Due to these ZnSe(Te) long time wasn't represent great interest for searcher and practical applications, because can't to be recommend for used in range X-ray with energy higher 120 -140 keV.

Situation changed on the contrary after developed dual energy X-ray inspection scanner, CT and digital radiography at whole. The theory and experiment show, that now it is best material for low energy array detectors [12-14]. This scintillator wide used in modern X-ray inspection system produce by Smith-Heimann firm [20].

REFERENCES

1. Thomas D.G. "A review of radiative recombination at isoelectronic donors and acceptors", *J. Phys. Soc. Japan.* (1966.) **21**. (Suppl.).pp.265–271.

2. Pat. 3586856, 1971 (USA). Radiation detector using isoelectronic trap material. Brown W.L., Height B., Madden T.S.

3. Ryzhikov V.D., Chaikovskii E.F. "Development of new scintillation materials on the basis of A^{II}B^{VI} compounds with an isovalent activator", *Izvestiya AN SSSR, ser. Physics.* (1979) **43**, No.6 pp.1150–1154. (rush.)

4. Kulp B.A. "Displacement of the cadmium atom in single crystal CdS by electron bombardment ", *Phys. Rev.* (1962). **125**. pp.1865–1869.

5. Detweiler R.M., Kulp B.A. "Annealing of radiation damage in ZnSe" , *Phys. Rev.* (1966) **146** pp. 513–516.

6. Susa N., Watanabe H., Wada M. "Effects of annealing in Cd or S vapour on photoelectric properties of CdS-single crystals", *Jap. J. Appl. Phys.* (1976) **15** pp.2365–2370.

7. Watkins J.D. "Lattice defects in II-VI compounds", *Inst. Phys. Cont. Ser.* (1977) № **31**. Chapter 1. pp.95–111.

8. Pat. 77055, 2006 (Ukraine), Grinyov B.V., Ryzhikov V.D., Silin V.I.

9. Pat. 74998, 2006 (Ukraine), Starzinskiy N.G., Grinyov B.V., Ryzhikov V.D.

10. Baltramejunas R., Ryzhikov V., Gavrushin V. et al. "Luminescence and nonlinear spectroscopy of recombination center in ZnSe:Te crystals" , *J. Luminescence.* (1992). **52.** pp. 71 – 81.

11. Baltrameunas R., Gavrushin V., Ryzhikov V., G.Raciukaitis. "Centers of radiative and nonradiative recombination in isoelectronically doped ZnSe:Te crystals", *Phys. B.* (1993) **185**. pp.245–249.

12. Grinyov B.V., Ryzhikov V.D., Seminozhenko V.P. "Scintillation detectors and systems of radiation monitoring on their base" (book). *K. : Akademperiodyka* (2010), 342p.

13. Ryzhikov V., Chernikov V., Galchinetskii L., Galkin S., Lisetskaya E., Opolonin A., Volkov V. "The use of semiconductor scintillation crystals $A^{II}B^{VI}$ in radiation instruments", *Journ. Cryst. Growth.* (1999). **197** pp. 655 – 658.

14. Ryzhikov V., Opolonin A., Pashko P., Svisch V. "Instruments and detectors based of scintillator crystal ZnSe(Te)", *Nuclear Instr. Meth.* (2005) **A-537**. pp.424–430.

15. Kulakov M.P., Fadeev A.V. "About stoichiometry crystals ZnSe were growing from melting", Izvestia Ac. Sc. USSR, *Ser. Inorg. Math.* (1981). **17** No.9. pp.1565-1570.

16. Fistul V.I. "Distribution isovalent dopant at sublattice multicomponent solid solution $A^{II}B^{VI}$ ", *Physica and Technique Semiconductors* (1983). **17** No.6. pp.1107 - 1110 (rush.)

17. Dmitriev Yu. N., Ryzhikov V.D. "About radiation hardness ZnSe(Te) crystals", *J. Atomic Energy* (1991). **70** No.2. pp.119-121 (rush.).

18. Ryzhikov V., Tamulaitis G., Starzhinskiy N. « Luminescence dynamics in ZnSe:Te scintillators », *J. Luminescence* (2003) **101**. pp.45–53.

19. Ryzhikov V., Starzhinskiy N. "Property and application isovalent doped $A^{II}B^{VI}$ compound based scintillators", J. Korea *Association for Radiation Protection.* (2005) **30**. pp.77–84.

20. Pat. US6445765B1, 2002 (USA). X-ray detecting apparatus. A. Frank, P. Schall, G. Geus.

Mater. Res. Soc. Symp. Proc. Vol. 1341 © 2011 Materials Research Society
DOI: 10.1557/opl.2011.1483

Nonlinear quenching rates in SrI₂ and CsI scintillator hosts

Joel Q. Grim, Qi Li, K.B. Ucer, R.T. Williams, *Wake Forest University*
A. Burger, P. Bhattacharya, E. Tupitsyn, *Fisk University*
G. A. Bizarri, W.W. Moses, *Lawrence Berkeley National Laboratory*

ABSTRACT

Using 0.5 ps pulses of 5.9 eV light to excite electron-hole concentrations varied up to 2×10^{20} e-h/cm³ corresponding to energy deposition within electron tracks, we measure dipole-dipole quenching rate constants K_2 in SrI₂ and CsI. We previously reported determination of K_2 directly from the time dependence of quenched STE luminescence in CsI. The nonlinear quenching rate decreases rapidly within a few tens of picoseconds as the host excitation density drops below the Förster threshold. In the present work, we measure the dependence of integrated light yield on excitation density in the activated scintillators SrI₂:Eu²⁺ and CsI:Tl⁺. The "z-scan" method of yield *vs.* irradiance is applicable to a wider range of materials, e.g. when the quenching population is not the main light-emitting population. Furthermore, because of using an integrating sphere and photomultiplier for light detection, the signal-to-noise is substantially better than the time-resolved method using a streak camera. As a result, both 2ⁿᵈ and 3ʳᵈ orders of quenching (dipole-dipole and Auger) can be distinguished. Detailed comparison of SrI₂ and CsI is of fundamental importance to help understand why SrI₂ achieves substantially better proportionality than CsI in scintillator applications. The laser measurements, in contrast to scintillation, allow evaluating the rate constants of nonlinear quenching in a population which has small enough spatial gradient to suppress the effect of carrier diffusion.

INTRODUCTION

Scintillators used as gamma-ray spectrometers can provide element and isotope-specific identification of material, as well as imaging or spatial localization of the information when used in suitable pixilated configurations. The essential function of the scintillator in spectroscopy mode is to produce a photopeak of detectable low-energy photons whose number is proportional to the energy of the incoming gamma-ray. If the *light yield* (photons/MeV incident) is a constant, the scintillator material would have perfect intrinsic proportionality. But neither the generation of excited states, nor the subsequent photoelectron path which may branch randomly into lower-energy scattered electron "delta rays", is the same from event to event. But it is well known that the rate of linear energy deposition along the track is a strong function of the electron energy as it slows. The energy deposition rate starts small at high electron energy, and increases to a very high rate around 100 eV. This general behavior is well known to surface scientists, for example, where it accounts for the so-called "universal curve of electron mean free path in solids" that reaches a minimum of about 0.3 nm near 100 eV in almost all materials. In the scintillator context a statement of the same phenomenon is that near a track end, the linear ionization rate reaches a maximum of about 3 electron-hole pairs per nanometer. Suffice it to say that there is a large variation of local e-h excitation density along the primary electron track from beginning to end, and maximum densities at track end have been estimated to reach about 2×10^{20} e-h/cm^3. At such high excitation densities nonlinear quenching of the e-h pairs to the ground state can be expected, and indeed nonlinear quenching at different rates from beginning to end of the electron trajectories is widely regarded as a main cause of nonproportionality.

However, it is one thing to identify a general phenomenon (nonlinear quenching) as a main cause, and quite another to identify what material parameters are predictive of the nonlinear quenching from scintillator to scintillator, and why. To illustrate, consider an electron track modeled as a cylinder of radius r. Along the length of the track, the linear energy deposition rate is given locally by dE/dx. Within the cylindrical track model, the concentration of excitations is thus proportional to $(dE/dx)/\pi r^2$. The lowest two orders of nonlinear concentration-dependent quenching are 2nd order, corresponding to dipole-dipole quenching (Förster transfer), and 3rd order, corresponding to free-carrier Auger recombination.

$$\left.\frac{dn}{dt}\right|_2 = -K_2 n^2 \propto -K_2 \left(\frac{dE/dx}{r^2}\right)^2 \qquad (1)$$

$$\left.\frac{dn}{dt}\right|_3 = -K_3 n^3 \propto -K_3 \left(\frac{dE/dx}{r^2}\right)^3 \qquad (2)$$

Variations of the three parameters on the right side could in principle each have major effect on the nonlinear quenching rate. The linear energy deposition rate dE/dx appears to 2nd or 3rd power; and the time-dependent track radius r(t) appears to the 4th or 6th power! The nonlinear quenching processes are proportional to the rate constants K_2 and K_3.

The change of dE/dx along the track causes nonlinear quenching to change accordingly and is thus implicated in nonproportionality. The linear energy deposition at high electron energy is known to scale from material to material approximately as the average electron density, proportional in turn to the mass density. But does its variation *along* the track differ significantly from material to material in a way predictive of nonproportionality? A survey of electron yield curves (~nonproportionality) for 29 scintillators measured in the SLYNCI experiment [1] did not find a strong relationship between nonproportionality and mass density or material-to-material variations of dE/dx. Instead, it found a clear correlation with host crystal classification such as alkali halides, multivalent halides, complex oxides, silicates, etc. In this paper, we look specifically at the pair of CsI and SrI$_2$ scintillator hosts which are composed of similar elements and have roughly similar mass densities of 4.51 and 5.46 gm/cm^3, respectively. If anything, one might expect SrI$_2$ with higher mass density and thus higher dE/dx to exhibit more serious nonlinear quenching. But SrI$_2$:Eu has significantly better gamma resolution than CsI:Tl.

We next turn attention to K_2, the 2nd order nonlinear quenching rate constant. What is its value in these two materials? Is the large difference in proportionality due to a smaller K_2 in SrI$_2$? What about K_3? We have previously measured K_2 in CsI by directly fitting the time dependence of STE luminescence quenching between 0.5 ps and roughly

200 ps as a function of on-axis excitation density produced by band-gap excitation with a 0.5 ps laser pulse. [2,3] The resulting (time-dependent) value of $K_2(t)$ in CsI was 2.4 x 10^{-15} cm^{3}s$^{-1/2}$ ($t^{-1/2}$). At excitation densities believed typical of a track end, the majority of nonlinear quenching was completed in about 10 ps. The nonlinear dependence of light yield on excitation density was measured in both undoped CsI (STE luminescence) and Tl-doped CsI (Tl luminescence), and was found to be the same. On the other hand, the 2.4 µs lifetime of the Tl activator did not exhibit nonlinear quenching behavior. It was concluded that in CsI:Tl, nonlinear quenching takes place in the STE population before the excitation is passed to Tl$^+$. [3]

We now want to compare K_2 across a wide variety of scintillators beginning with SrI$_2$:Eu and CsI:Tl reported in this paper. However, the method of fitting the quenching time as employed for CsI [3] is not widely applicable because it requires measuring the decay time of luminescence from an identified precursor species such as the STE in CsI. The activator decay time does not display nonlinear dependence on excitation density according to Ref. [3]. Therefore in this work we are measuring the *yield* of scintillation light versus on-axis initial excitation density. Because an integrating sphere can be used to channel light to a photomultiplier, the signal-to-noise achieved is sufficient to allow distinguishing K_2 and K_3 of the host.

EXPERIMENT
The experimental method is illustrated in Fig. 1(a). A lens on a translation stage focuses 5.9-eV, 0.5 ps laser pulses to a Gaussian beam waist. Therefore by translating the lens,

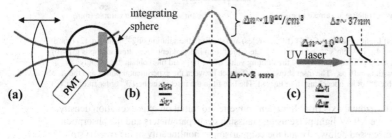

Figure 1. (a) Schematic of the experimental method with a translating lens focusing uv light on a sample and the subsequent luminescence channeled into a photomultiplier by an integrating sphere. (b) Illustration of the gradients of excitation density in the electron track of a scintillator in use, and (c) for laser interband excitation.

a Gaussian spot of variable radius but constant energy (photon number) may be produced at the front face of the sample. Thus, if there were no nonlinear quenching, the light output (registered by a photomultiplier coupled via an integrating sphere) should be constant during a "z-scan" of the lens translation.

Figures 1(b & c) illustrate that significantly different *gradients* of excitation density occur in the case of particle excitation of scintillators in actual use (e.g. SLYNCI measurements of electron yield) and in the case of laser excitation. Excitation densities up to 2 x 10^{20} e-h/cm^3 are produced in a track whose radius has been estimated by Vasil'ev to be 3 nm [4], producing a gradient of almost 10^{20} e-h/cm^3/nm. In such a gradient, carriers may diffuse out of the track before significant quenching occurs if they have sufficiently high mobilities [3]. In contrast, the penetration of the ultraviolet laser light is more than 30 nm, so for the same excitation density, the carrier gradient is about 10 times lower. Therefore, the laser experiment being reported gives information on nonlinear quenching rates themselves, mostly apart from the combination of nonlinear quenching and diffusion that determines light yield under particle excitation.

RESULTS

Figure 2 compares the light yield from CsI:Tl (0.3%) and SrI$_2$:Eu(1%) measured at constant pulse energy as a function of the sample surface position relative to the beam waist.

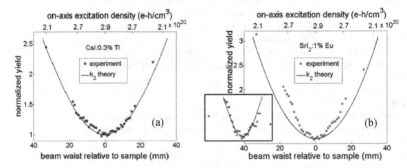

Figure 2. (a) Light yield from CsI:Tl (filled circles) and the fit to the solution of 2nd order Eq. (2) (solid line) using κ_2=2.4 x 10^{-15} cm^{3}s$^{-1/2}$ and t_R = 10 ps determined from independent decay-time methods [3]. (b) Light yield from SrI$_2$ (filled squares) and the limiting value of the fit (solid line) obtainable by only increasing κ_2 with t_R = 10 ps. The inset shows the difference between the experimental points and the steepest possible 2nd order quenching fit to Eq. (5). The line fit in the inset is the best 3rd order fit of Eq. (2).

The horizontal axis (top) has been converted to initial on-axis excitation density produced by the 5.9 eV light pulse using Gaussian beam parameters and the absorption coefficient discussed below. To aid the comparison of nonlinearity in the two crystals, the curves for both SrI$_2$:Eu and CsI:Tl are normalized so that the light yield is 1 at the beam waist. Assuming that dipole-dipole quenching is the dominant nonlinear process in iodide scintillators and that the nonlinear quenching is substantially completed in a time comparable to or less than the capture time on activators, then the nonlinear quenched fraction of the excited state population can be evaluated from Eq. (1), where K$_2$ has the explicit time dependence

$$K_2(t) = \kappa_2 \, t^{-1/2} \qquad (3)$$

for bimolecular decay of immobile species as discussed in Refs. [3, 5]. The time independent multiplier κ_2 will be used henceforth as the characterization of 2nd order quenching strength in a material. After integrating Eq. (1), we rearrange to give the quenched fraction QF:

$$QF = \frac{n_0 - n(t_R)}{n_0} = 2\kappa_2 n(t_R)\sqrt{t_R} \qquad (4)$$

where $n(t_R)$ is the carrier density remaining at time t_R when nonlinear quenching becomes negligible relative to radiative emission. We take this to be the average time for transfer from the system of free carriers and excitons to the activators. [6]

The left two terms can be solved for $n(t_R) = n_0(1-QF)$ and substituted into the right side, allowing to extract

$$\kappa_2 = \frac{QF}{2n_0(1-QF)\sqrt{t_R}} \qquad (5)$$

If we can establish the time t_R at which nonlinear quenching ceases to be significant, κ_2 can be extracted from fitting QF vs n_0, where normalized light yield is (1-QF).

There is experimental support for $t_R \approx 10$ ps in CsI:Tl. [3,6] Similar investigation of self-trapped exciton, self-trapped hole, and trapped charge populations on a ps time scale is needed in SrI$_2$:Eu and other materials to pin down a suitable value of t_R, the transfer time to activators. Furthermore, we have not yet been able to prepare and measure absorption in a ~80 nm film of hygroscopic SrI$_2$. To proceed with the comparison, we will assume for the time being that the 5.9 eV absorption coefficient in SrI$_2$ is the same as in CsI. Assumption of the same $t_R = 10$ ps and $\alpha(5.9 \text{ eV}) = 2.7 \times 10^5$ cm^{-1} for SrI$_2$ as in the better-studied CsI:Tl leads to the preliminary comparison

$\kappa_2 = 2.4 \times 10^{-15}$ cm^{3}s$^{-1/2}$ and κ_3 is negligible in CsI:Tl excited at 5.9 eV.
$\kappa_2 \geq 2.4 \times 10^{-15}$ cm^{3}s$^{-1/2}$ and $\kappa_3 \approx 1.6 \times 10^{-31}$ cm^{6}s^{-1} in SrI$_2$:Eu

DISCUSSION

First, we note that the value $\kappa_2 = 2.4 \times 10^{-15}$ cm^{3}s$^{-1/2}$ measured by decay-time methods in CsI [3] also gives a very good fit of the independent yield measurements in CsI:Tl in the present work, as shown in Fig. 2(a).

The value $t_R = 10$ ps is also in agreement with available time-resolved data [3, 6]. The fit is good enough that there is no need to invoke a 3rd order component of quenching in CsI:Tl within the measured range of excitation density. On the other hand, the data for SrI$_2$:Eu could not be fit to a similar degree with any choices of κ_2 and t_R. Upon plotting a family of QF vs n_0 curves from Eq. (5) for increasing κ_2, the steepness of the curve approaches a limiting value before experimental data could be matched. The difference between experiment and the maximum compatible 2nd order quenching was fit to the solution of the 3rd order Eq. (2), [see inset of Fig. 2(b)] yielding an estimate for κ_3. With further refinement, this could be quite important for understanding the different importance of dipole-dipole and Auger quenching in scintillators.

The present experiment was undertaken partly to see whether the better proportionality of SrI$_2$ relative to CsI might be attributable simply to a smaller rate constant κ_2 of 2nd order quenching in SrI$_2$. A firm answer still awaits measurement of the actual absorption coefficient of 5.9-eV light in SrI$_2$, but the comparison in Fig. 2 indicates that the relative value of κ_2 seems *not* to be in the direction that could be responsible for the main difference in proportionality of CsI and SrI2.

We now examine the third parameter on the right side of Eq. (1), namely the track radius r. Because of the 4th-power dependence of dipole-dipole quenching on r, it would not take much of a material-dependent change to have a substantial effect on nonlinear quenching. We have previously presented numerical modeling of diffusion and nonlinear quenching in scintillators, finding that the nonlinear quenched fraction depends on the ability of electrons and holes as well as excitons to diffuse out of the dense track core before nonlinear quenching can occur. [3, 7] The better proportionality of semiconductor detectors versus scintillators is well described by how mobile the electrons and holes are in ambipolar diffusion. In addition, the ratio of electron and hole mobilities was shown to govern the probability of pairing of carriers to form excitons in the extreme excitation gradient of particle tracks in a scintillator. [7] This control of the branching between free carriers and excitons governs the effective diffusion coefficient of the mixture of excitations in the track. The diffusion model was shown to give a quantitative fit of the nonproportionality parameter σ_{NP} cataloged for oxide and selenide scintillators by Setyawan et al [8], along with similar characterizations of semiconductor detectors [7]. Although halide scintillator proportionality has not yet been fit by the diffusion model [7] to the high precision achieved for oxides and semiconductors, the model nevertheless reproduces the familiar "halide hump" in simulated local light yield, appearing qualitatively similar to the actual halide hump exhibited in the experimental electron yield curves [1] for activated alkali halides and more weakly in other halides. [7]

Of the three parameters appearing on the right hand side of Eq. (1), variation of the track radius according to diffusion coefficients of the excitations in different materials appears at this point to provide the best documented link between independently measureable or calculable material parameters and nonlinear quenching (~nonproportionality).

*supported by the Office of Nonproliferaton Research and Development (NA-22) of the U. S. DOE, under contracts DE-NA0001012, DE-AC02-05CH11231 and DE-NA0000668.

REFERENCES

[1] S. A. Payne, W. W. Moses, S. Sheets, L. Ahle, N. J. Cherepy, B. Sturm, S. Dazeley, private communication of manuscript to be published (2011); S. A. Payne, N. J. Cherepy, G. Hull, J. D. Valentine, W. W. Moses, and W.-S. Choong, IEEE Trans. Nucl. Sci. **56**, 2506 (2009)

[2] Joel Q. Grim, Qi Li, K. B. Ucer, R. T. Williams, and W. W. Moses, Nucl. Instrum. Methods Phys. Res. A ; published online (2010), doi:10.1016/j.nima.2010.07.075.

[3] R. T. Williams, J. Q. Grim, Qi Li, K. B. Ucer, and W. W. Moses, Phys. Status Solidi B, **248**, 426 (2011).

[4] G. Bizarri, W.W. Moses, J. Singh, A.N. Vasil'ev, and R.T. Williams, J. Appl. Phys. **105**, 044507-0441 (2009).

[5] V. Nagirnyi, S. Dolgov, R. Grigonis, M. Kirm, L.L. Nagornaya, V. Savikhin, V. Sirutkaitis, S. Vielhauer, A. Vasil'ev, IEEE Trans. Nucl. Sci. **57**, 1182 (2010).

[6] R. T. Williams, K. B. Ucer, Joel, Q. Grim, K. C. Lipke, L. M. Trefilova, and W. W. Moses, IEEE Trans. Nucl. Sci. **57**, 1187 (2010).

[7] Qi Li, Joel Q. Grim, R.T. Williams, G. A. Bizarri, and W. W. Moses, J. Appl. Phys. **109**, 123716 (2011); doi:10.1063/1.3600070

[8] W. Setyawan, R. M. Gaume, R. S. Feigelson, and S. Curtarolo, IEEE Trans. Nucl. Sci., **56,** 2989 (2009).

Mater. Res. Soc. Symp. Proc. Vol. 1341 © 2011 Materials Research Society
DOI: 10.1557/opl.2011.1275

Bridgman Growth of SrI$_2$

Leonard Alaribe[1], Christian Disch[1], Alex Fauler[1], Ralf Engels[2], Egbert Keller[3], Angelica Cecilia[4], Tomy dos Santos Rolo[4], Michael Fiederle[1]

[1]FMF-Freiburger Materialforschungszentrum, Stefan-Meier-Straße 21, D-79104 Freiburg
[2]Forschungszentrum Jülich, Wilhelm-Johnen-Straße, 52428 Jülich
[3]Kristallographisches Institut, Hermann-Herder-Str. 5, D-79104 Freiburg
[4]Institut für Synchrotronstrahlung-ANKA, Forschungszentrum Karlsruhe

ABSTRACT

Eu^{2+}– activated strontium iodide is a promising material for x-ray and gamma ray detector. A lot of difficulties are though encountered growing strontium iodide crystals due to the high oxygen-sensitivity, hygroscopic property and high impurity concentration. Single crystals of SrI$_2$:Eu were grown from zone refined starting materials in silica ampoules. The crystals showed good optical qualities. The light yield of two samples cut from the same ingot was determined to be 53 000 photons/MeV and 119±22 photons/keV for a 0.4 cm^3 sample and a 360 μm sample respectively, indicating some level of light trapping in the bulk sample.
Keywords: scintillator, Bridgman growth method, energy resolution.

INTRODUCTION

The role of scintillators is fast expanding in many sectors of industrial measuring systems, fundamental research and medical imaging, where a high resolution measurement of the spatial distribution of radiation intensity in tissues is very necessary. Another important application of scintillators is in homeland security, nuclear nonproliferation and national defense as radiation detectors. Here a system is required to identify radioactive sources and their gamma ray signature. The identification of the isotope giving rise to a radiation is important to obtain information about the radiation type and radiation energy. Required for these applications is a scintillator with a relative high light yield, a good energy resolution, a fast scintillation decay time and a high density and high effective atomic number for a high gamma energy photoelectric absorption ($\propto \rho Z_{eff}^{3-4}$).

In this paper we report on the zone refinement, growth of SrI$_2$ single crystals with different Eu^{2+} concentrations, preparation and performance of a 6% Eu^{2+} doped SrI$_2$ in 0.4 cm^3 and 360 μm samples.

ZONE REFINING AND CRYSTAL GROWTH

Single crystals of 4N-purity SrI_2 (Alfa Aesar) doped with 3N-purity EuI_2 (Sigma-Aldrich) were grown with the Bridgman growth method. SrI_2 crystallizes in the orthorhombic crystal system, space group Pbca [1]. The density of the SrI_2 was given by the producer to be 4.55 g/cm^3.

SrI_2 is hygroscopic, so difference scanning calorimeter and thermo gravimetric methods were used to determine the level of hydration, the melting point and the best dehydration temperature of the charges prior to crystal growth. Fig.1 shows the melting point of the SrI_2 determined to be 512.6 °C and the dehydration temperature approximately 350 °C. Fig. 2 shows the melting point the EuI_2 at 534.9 °C.

While dehydrating a charge for the crystal growth, we noticed dark spots in the SrI_2 and decided to perform a zone refining on the starting material, in order to obtain better quality crystals after the crystal growth. The zone refining was done by loading the undoped SrI_2 in a silica ampoule in a glovebox purged with argon gas. The ampoule was closed with a vacuum valve and subsequently connected to a vacuum pump. The ampoule was evacuated to approximately 10^{-7} Torr and at the same time heated to approximately 350 °C to remove any moisture in material. After several hours, the ampoule was sealed and the zone refining done by melting the SrI_2-charge and lowering it through the temperature gradient of the Bridgman furnace at the rate of 1mm/hr.

In argon gas purged glovebox (moisture concentration: 0.522 ppm), the clean zone of the SrI_2 was cut off from the rest. For the crystal growth, a charge was weighed, doped and loaded in a new silica ampoule. The ampoule was evacuated to approximately 10^{-7} Torr and at the same time heated to approximately 350 °C for few hours to remove any residual moisture. The charge was heated to 570 °C and lowered through the temperature gradient of the Bridgman furnace at the rate of 0.5 mm/hr. For perfect crystal seeding, we provided the ampoules with 3 cm long and 3 mm wide capillaries.

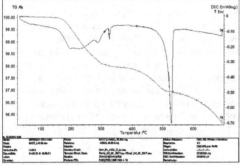

Fig.1: DSC and TG of SrI_2 showing the melting point and the possible dehydration temperature

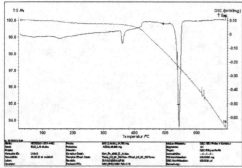

Fig.2: DSC and TG of EuI₂ showing the melting point

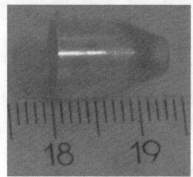

Fig.3: A 6% Eu doped SrI₂ single crystal harvested before polishing

SAMPLES PREPARATION

All preparation work was done in a glovebox purged with argon gas (moisture concentration 0.522 ppm). Crystal saw cutting was done using 0.3 mm diameter (well) diamond wire in paraffin oil. The crystals were ground with Buehler and Struers SiC grinding papers (p400, p800, p1200 and p4000). Crystals were polished using 1 µm (1 µGrit) Al₂O₃ (Pieplow & Brandt) in paraffin oil on a Buehler Texmet lap. At the end crystals were washed with kerosene.

For the measurement, two samples cut from the same ingot were prepared. Fig.4 shows the sample 1, an Ø10mm x 5mm crystal. The sample 1 was wrapped in a Teflon tape and hermetically encapsulated in an aluminum container with a quartz glass window for measurements. A sample was also prepared for x-ray imaging measurements. Fig.5 shows the sample 2 with a thickness of 360 µm glued to an optically polished aluminum disc using crystalbond 509. The sample 2 was covered with a quartz glass slide (150-170 µm thick) using apiezon wax W40. The water permeability of the apiezon wax was given by the producer to be 1.6×10^{-8} g/cm/hr/mm Hg at 25 °C.

Fig.4: Ø10mm x 5mm crystal (sample 1)

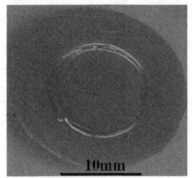

Fig.5: 360 micron thick sample 2

PERFORMANCE

To check for the performance of the sample 1 the pulse-height spectrum was recorded with the NIM Model 571 spectroscopy amplifier from Ortec using a shaping time of 1μs for the ^{137}Cs source (661 KeV). The light yield was determined by comparing the current pulse height of sample 1 (SrI$_2$:6%Eu, size: 0.4 cm^3) to that of a commercial NaI scintillator (size: 36.6 cm^3) reference sample and estimated to be 53 000 ph/MeV. The energy resolution was calculated to be 16.5% at 662 KeV for the full width at half maximum (FWHM).

We used x-ray imaging techniques to acquire a flat field image of the sample 2 and also estimate the light yield. The crystal was integrated in an X-ray imaging setup composed of an Optique Peter monochromatic microscope equipped with an objective (PLAPON 1.25x/NA 0.04) and a 2.5x eye-piece. According to the Rayleigh criterion, the resolution limit was 2.88 x 2.88μm^2 and the total magnification of the microscope was equal to 3.125x. The light yield of sample 2 (SrI$_2$:6%Eu, thickness: 360 μm) was estimated by comparing the flat field image to that of a YAG reference sample at 25 keV to be 119±22 ph/keV. Fig. 7 shows the flat field images of the SrI$_2$ on the left and that of the reference YAG sample on the right. The light yield was estimated

24

from the mean value of the counts measured in the same region of the flat field image of the scintillators. Broken quartz glass slide and stains on the glass are responsible for the white lines and dots on the SrI$_2$ flat field image. A tee mesh image was acquired with the SrI$_2$ showing a relatively good resolution fig.8.

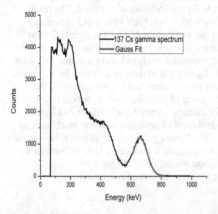

Fig.6: Gamma ray spectrum of the sample 1(SrI$_2$:6%Eu)

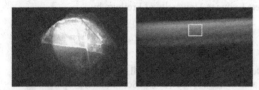

Fig.7: Flat field images of the SrI$_2$ (left) and YAG (right side) at 25 keV.

Fig.8: SrI$_2$:6%Eu^{2+} image tee mesh

DISCUSSION AND CONCLUSION

We grew Eu-activated SrI_2 single crystals with the Bridgman growth method. The crystals showed good optical qualities as a result of the zone refinement prior to the crystal growth. The light yield was determined for two samples cut from the same ingot to be 53000 ph/MeV for the 0.4 cm^3 sample 1 and 119 ± 22 ph/keV for the 360 micron sample 2. The results of the bulk sample 1 lies below literature values of over 90000 ph/MeV light yield and energy resolution less than 3% at 662 keV [2] [4] [5]. The light yield estimated for the sample 2 from the flat field images corresponds with literatures values. The low light yield of sample 1 could be attributed to some factors like structural defects and impurities in crystal, light trapping in crystal due to reflections at the scintillator detector window (quartz glass) interface, caused by high ratio of refractive indices between the crystal and the outer environment and the reflectivity of the packaging materials. Studies have shown that the coupling of scintillators with optical couplers to the detector window increases the amount of light extracted from the scintillator and its lack introduces insoluble problems with defining a standard level of polishing of the contact surfaces as well as the pressure force [6]. The measured crystals showed good optical qualities. No optical couplers were employed during the measurements, meaning their lack could have influenced the amount of light extracted from the scintillator. The good resolution of the tee mesh image acquired with the SrI_2 makes it a possible candidate for x-ray tomography applications.

ACKNOWLEDGMENTS

This scientific work was funded by the Albert-Ludwigs-University of Freiburg by 'Anschubfinanzierung BIRD'.

REFERENCES

1. Edgar V. van Loef, Cody M. Wilson,   Nerine J. Cherepy, Giulia Hull, Stephen A. Payne, Woon-Seng Choong, William W. Moses, Kanai S. Shah " Growth and Scintillation Properties of Strontium Iodide Scintillators," *IEEE Trans. Nucl. Sci.*, VOL. 56, NO. 3, JUNE 2009
2. Nerine Cherepy, L. Boatner, A. Burger, K. Shah, Steve Payne, Alan Janos, Austin Kuhn: Overview of strontium iodide scintillator materials. Llnl Presentation 1 Apr. 2010
3. J. B. Birks: The Theory and Practice of Scintillation Counting. Pergamon Press 1964
4. Payne et al.: BARIUM IODIDE AND STRONTIUM IODIDE CRYSTALS AND SCINTILLATORS IMPLEMENTING THE SAME, United States Patent Application Publication [Appl. No.: 12\255,375]
5. N.J. Cherepy, S.A. Payne, S.J. Asztalos, G. Hull, J.D. Kuntz, T. Niedermayr, S. Pimputkar, J.J. Roberts, R.D. Sanner, T.M. Tillotson, E. van Loef, C.M. Wilson, K.S. Shah, U.N. Roy, R. Hawrami, A. Burger, L.A. Boatner, W. S. Choong, W.W. Moses: Scintillators with Potential to Supersede Lanthanum Bromide
6. Michał Gierlik, Marek Moszyn´ski, Antoni Nassalski, Agnieszka Syntfeld-Ka´zuch, Tomasz Szcz˛es´niak, LukaszS´ widerski: Investigation of Absolute Light Output Measurement Techniques, IEEE Trans. Nucl. Sci., Vol. 54, No. 4, Aug. 2007
7. Piotr A. Rodnyi: Physical process in inorganic scintillators, CRC Press LLC 1997

CdZnTe

Mater. Res. Soc. Symp. Proc. Vol. 1341 © 2011 Materials Research Society
DOI: 10.1557/opl.2011.1478

Growth of detector-grade CZT by Traveling Heater Method (THM): An advancement

U. N. Roy[1], S. Weiler[1], J. Stein[1], M. Groza[2], A. Burger[2], A. E. Bolotnikov[3], G. S. Camarda[3], A. Hossain[3], G. Yang[3] and R. B. James[3]

[1]FLIR Radiation Inc., 100 Midland Road, Oak Ridge, TN 37830
[2]Department of Physics, 1000, 17th Avenue North, Fisk University, Nashville, TN 37208
[3]Brookhaven National Laboratory, Upton, NY 11793

ABSTRACT

In this present work we report the growth of $Cd_{0.9}Zn_{0.1}Te$ doped with In by a modified THM technique. It has been demonstrated that by controlling the microscopically flat growth interface, the size distribution and concentration of Te inclusions can be drastically reduced in the as-grown ingots. This results in as-grown detector-grade CZT by the THM technique. The three-dimensional size distribution and concentrations of Te inclusions/precipitations were studied. The size distributions of the Te precipitations/inclusions were observed to be below the 10-μm range with the total concentration less than 10^5 cm^{-3}. The relatively low value of Te inclusions/precipitations results in excellent charge transport properties of our as-grown samples. The $(\mu\tau)_e$ values for different as-grown samples varied between 6-20 $\times10^{-3}$ cm^2/V. The as-grown samples also showed fairly good detector response with resolution of ~1.5%, 2.7% and about 3.8% at 662 keV for quasi-hemispherical geometry for detector volumes of 0.18 cm^3, 1 cm^3 and 4.2 cm^3, respectively.

INTRODUCTION

In spite of continuous efforts to develop novel room-temperature detector materials, CdZnTe (CZT) remains to be the most promising semiconductor material for room-temperature nuclear detector applications for almost two decades. Presently there is an increasing demand for larger volume, especially larger thickness (>10 mm) CZT detectors for homeland security applications for fast and unambiguous nuclide identification. Thicker detectors provide sufficient stopping power for higher energy gammas and better standoff detection. Although the Travelling Heater Method (THM) technique is well established for the growth of large-volume CZT crystals, its main bottleneck until today has been the need for post-growth annealing for detector applications [1, 2]. The THM technique offers many advantages over melt-growth techniques. The main advantage is the fairly uniform Zn concentration along the growth direction [3, 4], which is essential for good charge transport especially for thick detectors in addition to better yield. As it is well known that THM is a lower temperature growth process compared to Bridgman (i.e., much below that the melting point of CZT), advantages of lower growth temperature are less or no chance of explosion of the growth ampoule, less contamination from the crucible, and less defect density. Schoenholz et al. [5] reported lower etch pit density for THM-grown CdTe compared to Bridgman-grown CdTe. They also demonstrated that the defect density reduces drastically in the grown crystal compared to the seed. Recently Yang et al. [6] demonstrated that

Te acts as an impurity gettering agent, this also is an added advantage of self-purification of the grown ingots from the Te-rich solvent by the THM technique. Thus, crystals with less thermal stress, less defects and higher purity with composition uniformity can be grown by a THM technique. The THM technique however has the main disadvantage of having Te-rich CZT and CZT interface. This interface is the cause for added Te inclusions in the grown ingots other than the formation of Te precipitations due to retrograde solubility. In order to minimize the formation of Te inclusions during growth from the interface, a microscopically flat and clean interface is essential to achieve as-grown detector-grade material. It is well known now that the Te inclusions/precipitations have detrimental effect on the device performance [7, 8]. The main purpose for post-growth annealing is the removal of relatively large (> 5-10-micron diameter) Te inclusions formed during growth. Minimization of Te inclusions originating at the interface during THM growth results in as-grown detector-grade material with good charge transport properties. The elimination of the post-growth annealing process will reduce production costs and improve the availability of large-size detectors. In this paper we will discuss the volumetric size distribution and concentration of Te precipitations/inclusions of as-grown CZT crystals and the minimization of Te inclusions by controlling the growth interface. Experimental data on charge transport properties, internal electric field estimation and detector performance of as-grown CZT samples will be presented and discussed.

EXPERIMENT

All the crystals were grown from as-received 6N purity CdZnTe from 5N Plus Inc. with the nominal concentration of 10% Zn. 6N purity Te from Alfa Aesar was used for the solvent, and 6N purity In was used as the dopant. The crystals were grown by the THM technique from a Te-rich solvent in conically tipped quartz ampoules. Prior to loading, the inner walls of the ampoules were coated with graphite by cracking high-purity acetone. The details of the growth procedure are discussed elsewhere [9].

The growth interfaces were investigated by cutting the wafers along the growth direction near the interface. The microscopic investigation of the interfaces was carried out on mirror-finished polished samples using a high magnification optical microscope (Zeiss Axio Imager M1m) after final polishing with a 0.05-μm alumina suspension. For IR transmission imaging of the full 52-mm diameter wafers, slabs were cut perpendicular to the growth direction and polished to mirror finish surface on both sides. An automated IR transmission microscopy system was used for mapping the Te inclusions/precipitates. The system comprises a large field-of-view microscope objective, a 3.5 x3.5 μm^2 pixel digital CCD camera that produces 2208x3000 pixel images, and a computer-controlled XYZ translation stage. Stacks of images were acquired, each focused at different depths in the wafers. The resolution of the IR imaging system is sufficient to quantify the Te inclusions/precipitates down to ~1 μm. We employed an iterative algorithm to identify the inclusions and evaluate their sizes. The experimental procedure is described in greater detail elsewhere [10].

Samples of different sizes were cut and polished from the as-grown ingots for charge transport measurements and detector fabrications. For charge transport measurements, planar gold electrodes were used by sputtering technique. A collimated [241]Am alpha source was used for the $(\mu\tau)_e$ measurements. The $(\mu\tau)_e$ values were estimated by plotting the charge collection efficiency versus applied voltage and then fitting to the Hecht equation. Different devices were fabricated with volumes ranging from ~0.2 cm^3 to over 4 cm^3 from the as-grown ingots. Most of the devices were tested with a quasi-hemispherical geometry; the geometry consists of a pixelated anode and five-side cathode. All the contacts were produced from sputtered gold.

RESULTS AND DISCUSSION

The grown ingots were easily removed from the quartz ampoules without any sticking problem, mainly due to the graphite coating on the inner wall of the crucibles. A thin coating of Te was observed around the grown ingots. We are routinely growing 52-mm diameter ingots by THM technique, and most of the results presented here will be from 52-mm diameter ingots unless otherwise mentioned. As discussed earlier, there are two obvious sources for formation of Te inclusions/precipitations. One is common for both THM- and melt-growth techniques, which is the retrograde solubility. The size and concentration of the Te precipitations due to retrograde solubility are supposed to be less in THM-grown ingots compared to melt-grown ingots due to the lower temperature growth. The second reason for forming Te inclusions in THM-grown crystals is the trapping of Te-rich CZT at the growth interface. The trapping can be reduced or minimized by controlling the growth interface to become more microscopically flat, which is the key factor for achieving as-grown detector grade CZT. In addition to that, the macroscopic shape of the interface plays an important role in the grain growth. Flat/convex interface results in growth of single crystalline/large grain crystals. We are continuously trying to improve the growth interface in order to achieve a microscopically and macroscopically flat interface to minimize Te inclusions due to trapping and to increase the grain size. Figure 1 shows a typical as-grown ingot cut along the growth direction. The shiny portion near the top of the ingot is the Te-rich CZT solvent portion, and the interface between the Te-rich CZT and grown CZT is reasonably flat as shown in Fig. 1.

   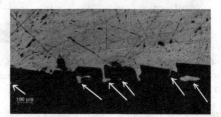

Fig. 1. 52-mm diameter CZT ingot by THM   Fig. 2. Morphology of the growth interface

Figure 2 represents an optical micrograph taken in a reflecting mode, the irregular shape of the interface depicting the trapping of Te-rich CZT in the as-grown ingot is clear. The lower dark portion is the grown CZT, and the bright portion is the Te-rich CZT solvent. The trapping of Te-rich CZT at the interface is evident from Fig. 2, as shown by the arrows. These regions lead to Te-rich inclusions in the as-grown part of the ingot. In order to reduce the trapped Te-rich volumes, it is essential to achieve a microscopically flat interface resulting in as-grown detector grade CZT that is relatively free of Te inclusions.

31

Fig. 3. Smooth growth interface

After optimization of our growth process, a microscopically flat interface could be achieved. Figure 3 is the typical representation of the microscopically flat interface and illustrates minimization of the trapped Te-rich CZT solution in the grown ingot. As shown in Fig. 2, the shapes of the trapped Te inclusions are highly irregular in nature. The trapped Te inclusions at the growth interface severely affect the yield and the device performance. Figure 4a shows the

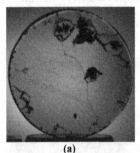
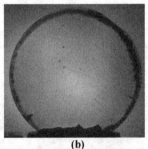

(a)   **(b)**

Fig. 4. IR transmission image of a 52-mm diameter CZT. a) Example of trapped Te inclusions at the interface during growth; b) fairly clean wafer resulting from a microscopically clean melt-solid interface.

IR transmission image of a 52-mm diameter full wafer grown with an irregular growth interface. The black portions are the trapped Te-rich CZT solvent inclusions severely affect the detector yield. It demonstrates the importance of controlling the interface during growth. Figure 4b shows the relatively clean as-grown CZT wafer by THM technique from a microscopically flat growth interface after optimization of the growth parameters. It is to be noted that in most of the ingots, Te inclusions were found to be more concentrated near the periphery of the ingots as shown in Figs. 4a and 4b.

The size distribution and concentration of Te inclusions/precipitations are critical parameters for device-grade CZT materials. Presence of these defects, especially with sizes of 3 μm and larger are known to severely degrade the device performance, especially for long-drift-length (thicker detector) devices. Higher concentrations and larger sizes of the Te inclusions/precipitations in the CZT material restrict the use of thicker detectors as gamma-ray spectrometers. In order to evaluate our as-grown CZT crystals, we have investigated the size distribution and concentration of Te inclusions/precipitations present in the bulk of the crystals. Figure 5a shows a typical 3D volumetric distribution of Te inclusions/precipitations over the volume of 1x1.5x3 mm³ for CZT grown using our THM method grown with optimized conditions. The total concentration of the Te inclusions/precipitations over the scanned volume was estimated to be 1.3×10^5 cm⁻³ and is comparable to the recently reported value for commercially available CZT samples [11]. The size distribution of the Te inclusions/precipitations over the same volume is shown in Fig. 5b.

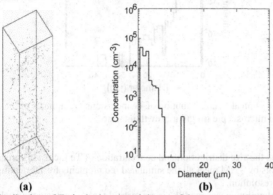

(a)   **(b)**

Fig. 5. a) 3D distribution of Te inclusions/precipitates of the as-grown CZT sample within a 1x1.5x3 mm³ volume; b) the size distribution over the same volume.

It is to be noted that the main histogram extends below 10 μm, and was consistent at different positions of the wafers and also for different ingots, while the distribution was recently reported to be extend from 15-25 μm for many commercially available CZT samples [11]. In our ingots grown with a microscopically irregular interface, the main histogram of the size distribution of the Te inclusions/precipitations was observed to extend to over 25 μm as shown in Fig. 6. The total concentration observed was ~ 8×10^5 cm⁻³. It is also evident that most of the concentrations are around or below 3 μm as shown in Fig. 5b. The concentrations of Te precipitates/inclusions with a diameter of ~4 μm or larger was found to fall in the region of 6-8 $\times 10^3$ cm⁻³. Bolotnikov et al. [12] found such small inclusions (size ~3 μm) degrade the device's performance only when their concentrations are above 10^6-10^7 cm⁻³, while uniformly distributed inclusions below 1-μm size do not adversely affect performance. As shown in Fig. 5b, discrete occurrences of larger diameter Te precipitations /inclusions were also observed. In some places of the wafers even larger diameter up to 30-μm size inclusions were detected with a concentration of slightly higher than 100 cm⁻³; these might be formed due to trapping of Te-rich solvent at the interface during growth. The simulation study however shows that for 15-mm thick detector, the resolution should be less than 2% at 662 keV for concentration of 10^3 cm⁻³ for 20-μm diameter size Te

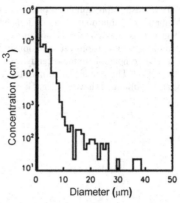

Fig. 6. Typical size distribution of Te inclusions/precipitates of the ingot grown with a microscopic irregular growth interface.

inclusions [12]. Thus, the size distribution and concentration of Te inclusions/precipitates in our as-grown CZT crystals by THM meet all the simulated requirements for fabricating 15-mm-thick detectors with good resolution.

Low temperature photoluminescence (PL) is a powerful tool to evaluate the quality of the material. Figure 7 shows the typical PL spectrum at 5 K for as-grown CZT by the THM technique. The sample was etched in 2% bromine-methanol solution prior to the measurement. The spectrum consists of three main regions; they are: i) near band edge region, ii) donor-acceptor region, and iii) defect band, generally known as the A-center. The spectrum consists of a dominant donor-bound exciton (D^0, X) peak centered at 1.653 eV; the ground state free exciton peak (X_1) and the upper polariton band (X_{up}) are visible as shown in Fig. 7b. The dominant (D^0, X) peak and the observed free excitonic peak are the reflection of high-quality detector-grade material [13,14]. The sharpness of the (D^0, X) peak also is an indicator of the quality of the material [13]. The FWHM of the (D^0, X) peak after Lorentzian fitting was found to be 2.2 meV and agrees well with the reported value of ~ 3 meV [14]. The acceptor-bound exciton (A^0, X) was observed at the peak energy of 1.64 eV. The peak centered at an energy 1.618 eV was assigned as the LO phonon replica of (A^0, X), which we designate as (A^0, X-1LO). The donor acceptor pair (DAP) transition was observed at 1.605 eV as shown in Fig. 7c. The LO phonon replicas of DAP are also visibly seen in Fig. 7c.

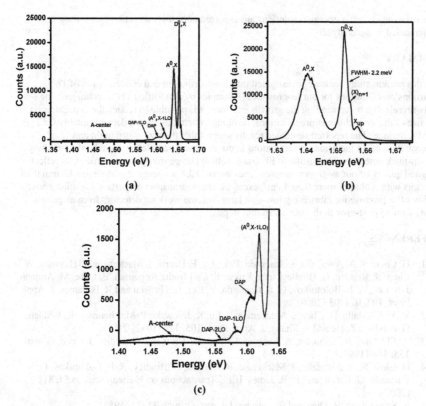

Fig. 7. PL spectra at 5 K of as-grown CZT by THM: a) full spectrum, b) near-band-edge region, and c) donor-acceptor and defect band regions.

The broad band centered at ~ 1.48 eV is the defect band, assigned as A-center; it agrees well with the literature [13, 15]. The A-center peak originates due to the Cd vacancy-In complex.

The charge transport properties of our as-grown CZT samples were investigated up to a thickness of 10 mm. The $(\mu\tau)_e$ values varied from 6×10^{-3} to 2×10^{-2} cm^2/V for different ingots, with the best mobility of 970 cm^2/Vs. The resistivities of the as-grown samples for different ingots were found to be consistently in the range of $(1-3)\times10^{10}$ ohm-cm. The internal electric fields were found to deviate from the ideal ones; they are comparable to those observed for commercially available CZT samples. Many detectors of different volumes were fabricated from the as-grown ingots. The as-grown samples showed fairly good detector response with resolutions of ~1.5%, 2.7% and 3.8% at 662 keV for a quasi-hemispherical geometry for detector volumes of 0.18 cm^3, 1 cm^3 and 4.2 cm^3, respectively. Here, no electron charge-loss corrections were used; much better energy resolutions are expected once this correction factor is implemented in the analysis.

SUMMARY

In this present study we have investigated the size distribution and concentration of Te inclusions/precipitations for our as-grown CZT samples by a modified THM technique. The study reveals that by controlling the growth interface, it is possible to reduce the concentration and size of the Te inclusions/precipitations, resulting in better as-grown detector quality CZT by THM technique. The reported simulated results suggest that the size distribution and concentration of Te inclusions/precipitations in our as-grown sample are capable of fabricating 15-mm-thick detectors. The results of PL and excellent charge transport properties also reflect the good quality of our as-grown samples. The observed 3.8% energy resolution for 16-mm thick detectors with volumes more than 4 cm^3, excellent charge transport properties, and high crystal quality offer promise for fabricating low-cost large-volume working detectors from as-grown ingots with no post-growth thermal annealing steps.

REFERENCES

1. H. Chen, S. A. Awadalla, K. Iniewski, P. H. Lu, F. Harris, J. Mackenzie, T. Hasanen, W. Chen, R. Redden, G. Bindley, Irfan Kuvvetli, Carl Budtz Jørgensen, P. Luke, M. Amman, J. S. Lee, A. E. Bolotnikov, G. S. Camarda, Y. Cui, A. Hossain and R. B. James, J. Appl. Phys. **103**, 014903 (2008).
2. S. A. Awadalla, H. Chen, J. Mackenzie, P. Lu, K. Iniewski, P. Marthandam, R. Redden, G. Bindley, Z. He and F. Zhang, J. Appl. Phys. **105**, 114910 (2009).
3. A. El Mokri, R. Triboulet, A. Lusson, A. Tromnos-Carli and G. Didier, J. Cryst. Growth **138**, 168 (1994).
4. H. Chen, S. A. Awadalla, J. Mackenzie, R. Redden, G. Bindley, A. E. Bolotnikov, G. S. Camarda, G. Carini and R. B. James, IEEE Transactions on Nuclear Science **54**, 811 (2007).
5. R. Schoenholz, R. Dian and R. Nitsche, J. Cryst. Growth **72**, 72 (1985).
6. G. Yang, A. E. Bolotnikov, Y. Cui, G. S. Camarda, A. Hossain and R. B. James, J. Cryst. Growth **311**, 99 (2008).
7. A. E. Bolotnikov, G. S. Camarda, G. A. Carnini, Y. Cui, L. Li and R. B. James, Nucl. Instr. Meth. Phys. Res. **A 571**, 687 (2007).
8. D. S. Bale, J. Appl. Phys. **107**, 014519 (2010).
9. U. N. Roy, A. Gueorguiev, S. Weiler and J. Stein, J. Cryst. Growth **312**, 33 (2010).
10. A. E. Bolotnikov, N. A-Jabbar, O. S. Babalola, G. S. Camarda, Y. Cui, A. Hossain, E. M. Jackson, H. C. Jackson, J. A. James, K. T. Kohman, A. L. Luryi and R. B. James, IEEE Trans. on Nucl. Sci. **55**, 2757 (2008).
11. A. E. Bolotnikov, O. S. Babalola, G. S. Camarda, Y. Cui, S. U. Egarievwe, R. Hawrami, A. Hossain, G. Yang and R. B. James, IEEE Trans. on Nucl. Science **57**, 910 (2010).
12. A. E. Bolotnikov, G. S. Camarda, G. A. Carini, Y. Cui, K. T. Kohman, L. Li, M. B. Salomon and R. B. James, IEEE Trans. on Nucl. Science **54**, 821 (2007).

13. J. E. Toney, B. A. Brunett, T. E. Schlesinger, J. M. Van Scyoc, R. B. James, M. Schieber, M. Goorsky, H. Yoon, E. Eissler, and C. Johnson, Nucl. Instrum. Methods **A 380**, 132 (1996).
14. T. E. Schlesinger, J. E. Toney, H. Yoon, E. Y. Lee, B. A. Brunett, L. Franks and R. B. James, Materials Sc. and Engr. Reports **32,** 103 (2001).
15. G. Yang, W. Jie, Q. Li, T. Wang, G. Li and H. Hua, J. Cryst. Growth **283,** 431 (2005).

Mater. Res. Soc. Symp. Proc. Vol. 1341 © 2011 Materials Research Society
DOI: 10.1557/opl.2011.1479

Low Temperature Crystal Growth and Characterization of $Cd_{0.9}Zn_{0.1}Te$ for Radiation Detection Applications

Ramesh M. Krishna[1], Timothy C. Hayes[1], Peter G. Muzykov[1], and Krishna C. Mandal[1]

[1]Department of Electrical Engineering, University of South Carolina, Columbia, SC 29208, USA

ABSTRACT

$Cd_{0.9}Zn_{0.1}Te$ (CZT) detector grade crystals were grown from zone refined Cd, Zn, and Te (7N) precursor materials, using the tellurium solvent method. These crystals were grown using a high temperature vertical furnace designed and installed in our laboratory. The furnace is capable of growing up to 8" diameter crystals, and custom pulling and ampoule rotation functions using custom electronics were furnished for this setup. CZT crystals were grown using excess Te as a solvent with growth temperatures lower than the melting temperatures of CZT (1092°C). Tellurium inclusions were characterized through IR transmittance maps for the grown CZT ingots. The crystals from the grown ingots were processed and characterized using I-V measurements for electrical resistivity, thermally stimulated current (TSC), and electron beam induced current (EBIC). Pulse height spectra (PHS) measurements were carried out using a ^{241}Am (59.6 keV) radiation source, and an energy resolution of ~4.2% FWHM was obtained. Our investigation demonstrates high quality detector grade CZT crystals growth using this low temperature solvent method.

INTRODUCTION

Among the wide variety of semiconductor materials available today, CdZnTe (CZT) is one of the most promising materials for producing nuclear radiation detectors. Unlike other popular options on the market, CZT operates at room temperature (300K), has a high average atomic number (Z = ~50), a wide band-gap (≥1.5 eV at 300K), and has a high density (5.8 g/cm^3) [1]. Applications for nuclear radiation detectors, such as national security, medical imaging, and high-energy astrophysics have increased the need for spectrometer-grade detectors, which has driven the commercial development of CZT in recent years.

However, one major impediment to CZT's widespread use is the high cost of producing large defect-free bulk single crystals of CZT. This is due to the presence of Te inclusions in the bulk crystals, Zn segregation, and defects in the crystal structure caused by impurities and mechanical stress. Furthermore, the high melting temperature of CZT increases the risk of contamination from impurities present in the fused silica ampoules [2].

A low temperature growth method has been developed and implemented for producing large volume single crystals of CZT, using Te-solvent method. Growth temperatures are kept below the melting point of CZT, reducing the risk of contamination from the quartz ampoules. Details of the crystal growth, characterization, detector fabrication, and detector performance are presented in the following sections.

EXPERIMENT

Crystal Growth of CZT ($Cd_{0.9}Zn_{0.1}Te$:In; 5 to 25 ppm In) was performed using in-house zone refined (~7N) Cd, Zn, and Te precursor materials. The zone refined precursors were prepared in sealed quartz ampoules placed in a tube furnace mounted on a track actuator moving at

Figure 1: Photo of ingots produced using the low temperature growth method.

30mm/day. The refiner makes 40-45 passes into the furnace, taking 30-35 days for each precursor. Glow discharge mass spectroscopy (GDMS) analysis confirmed our in-house zone refined (ZR) precursor materials (Cd, Zn and Te) are of 7N purity. Typical charges of CZT precursors had 36:4:60 at% Cd:Zn:Te to yield 50 at% excess Te for crystal growth. The dopant and precursors were inserted into quartz ampoules of wall thickness ~4.0 mm and sealed under 10^{-6} torr vacuum, and loaded into a three-zone vertical growth furnace. The growth furnace used for this experiment was constructed in the laboratory at USC, and fitted with custom pulling and rotation mechanisms. In-house electronics and software were developed to control and automate all functions pertaining to crystal pulling and rotation. The furnace is capable of producing ingots with a diameter of 8", however for this experiment, ingots were produced in the range of 1-2" diameter. Growth was initiated by lowering the ampoules into a 965°C hot zone at a rate of 1-2 mm/hour. Typical temperature gradient was 3.9°C/cm, and the gradient at the crystal growth zone was 0.3°C/cm. Ampoules were also rotated at a rate of 12 rotations per hour. The grown CdTe and CZT crystals were then cut, wafered, polished (down to 0.05 μm alumina), and etched (using 2% Br_2-MeOH) in our semiconductor processing laboratory. After polishing the CdTe and CZT crystals to a mirror finish, samples were used for characterization and detector fabrication.

Characterization on the grown CZT crystals was performed to evaluate the optical and electrical qualities of the material, as well as the detection characteristics of the crystals. Several defect studies were conducted on the grown CZT, in order to fully characterize and understand CZT radiation detection performance. Thermally Stimulated Current (TSC) studies were performed on CZT in order to determine deep level defects within the crystals. TSC measurements were conducted in the temperature range of 94 – 400 K under a 2×10^{-5} Torr vacuum in a low temperature microprobe chamber (See schematic in Figure 4a) [3]. Tungsten-halogen white light was used at 94K to fill energy traps for this experiment.

Electron Beam Induced Current (EBIC) contrast images were taken of a fabricated CZT diode, and correlated to IR transmittance images of the CZT surface. EBIC imaging was performed using a JEOL-35 SEM (See schematic in Figure 2) [4]. A reverse bias

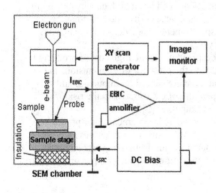

Figure 2: Schematic showing Electron Beam Induced Current (EBIC) setup.

voltage of 80V was applied to the CZT diode, and EBIC contrast images were taken on the Schottky contact surface.

In addition to defect studies, bulk I-V characteristics, bulk resistivity, and optical transmission/absorption studies are presented. After performing characterization on the grown crystals, 7x7x5 mm^3 CZT Frisch collar radiation detectors [6] were fabricated from suitable portions of the CZT ingots, and nuclear detection measurements were carried out using an ^{241}Am source. Single element gold contacts were fabricated to all detectors, and the cathode was irradiated through a small hole underneath the detector bias stage. Our detectors were biased using an Ortec 556 high voltage supply. The charge signal generated by the detectors was amplified using an Amptek A250CF preamplifier and an Ortec 671 spectroscopy amplifier, and read into a Canberra multi-channel analyzer.

RESULTS AND DISCUSSION

Prior to the crystal growth, GDMS data was acquired for the Cd, Zn, and Te precursor materials used for the crystal growth. The analysis showed a lower concentration of impurities than the detector was able to detect for the majority of impurity elements, and lower than 100 ppb for other trace elements such as silicon, copper, selenium, and oxygen. After performing the crystal growth, ingots from 1-2" diameter and 2-7" long have been produced (Figure 1). After preparing several single crystal wafers from the middle of the ingots, gold contacts were applied using electroless gold deposition using a gold chloride solution, and I-V characteristics were measured. Schottky contacts were fabricated by using indium metal as a Schottky contact on a CZT wafer of dimension 7x7x5 mm^3 and the leakage current at -1000V was determined to be less than 5nA (Figure 3a). Additionally, optical absorption and transmission results (Figure 3b) were obtained for a thin wafer of CZT, and the band gap was calculated to be 1.55 eV.

For TSC measurements, CZT samples were measured at various heat rates and various voltage bias conditions. TSC spectra were initially taken using heat rates of 4, 8, and 15 K/min, and the dark current was subtracted from the acquired plot. After subtracting the dark current, several distinct TSC peaks were observed. Further, TSC spectra were gathered for varying voltage bias (Figure 4b) and varying heat rates at 0V bias.

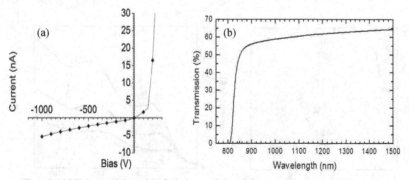

Figure 3: (a) I-V characteristics and (b) Optical transmission of a low temperature grown CZT crystal.

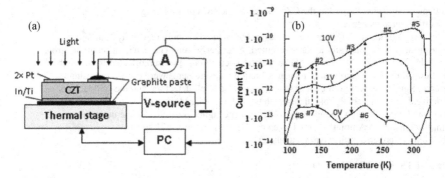

Figure 4: (a) Schematic showing TSC setup for CZT (b) TSC spectra for CZT for 15 K/min heat rate at 0, 1, and 10V.

An Arrhenius plot was generated for the TSC energy peaks, based on the following formula [3]:

$$E_T / kT = \ln(T_m^4 / \beta) + \ln(1 \times 10^{17} \, \sigma / E_T) \tag{1}$$

The activation energies were then calculated based on the Arrhenius plot [3], and are listed in Table 1. In addition to calculating the activation energy, the capture cross section was calculated from this information using previously reported methods [3] for the traps associated with the TSC peaks. One key observation made in this analysis is that as the bias voltage increases, the TSC peak intensity increases in proportion with the increase of the depletion region width and the number of traps contributing to the TSC signal. This confirms that the measured TSC spectra are related to traps in the bulk crystal rather than imperfections in the surface or semiconductor-metal interface. This suggests that the traps originated during the crystal growth process, as opposed to the crystal processing stages. Assignment of the TSC peaks to impurities and defects were performed based on existing work on CZT [5] [6]. Further analysis of these results is in progress and will be reported later.

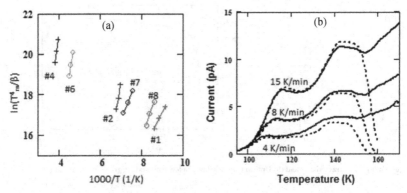

Figure 5: (a) Arrhenius plot from the TSC spectra (b) Theoretical fit of peaks #1 and #2 in Figure 4b.

Prior to the EBIC characterization, IR images were taken of the surface of the CZT crystals. Figure 6a shows an IR image of a small section of a CZT crystal. Te inclusions can be seen as black spots in the IR image. Pt (Ohmic) and In (Schottky) contacts were then fabricated to one of the grown CZT crystals. EBIC contrast images of the CZT crystal are shown in Figure 6b. The contrast image shows concentric dark and white

TSC peak #	T_m (K)	Activation energy (eV)	Capture cross section (cm^2)	Possible defect
1	116*	0.18	$10^{-19} - 10^{-16}$	$Vcd^{0/-}$
2	145*	0.46	10^{-12}	$(Tecd+2Vcd)^{2-/-}$
4	~261*	0.86	-	$(Tecd+2Vcd)^{3-/2-}$ or $(Tecd+Vcd)$
6	222**	0.58	4×10^{-14}	$Tecd^{+2/+}$
7	140**	0.26	10^{-18}	$Tecd^{0/+}$
8	121**	0.27	10^{-15}	$Vcd^{2-/-}$

* from TSC spectra at 15 K/min heat rate and 10 V bias
** from TSC spectra at 15 K/min heat rate and 0 V bias

Table 1: Results deduced from TSC measurements shown in Fig 4 and 5.

regions. Dark regions correspond to higher EBIC current, while lighter regions correspond to lower EBIC current. The morphology of the EBIC contrast image is similar with the IR transmission image morphology, with higher contrast regions in the contrast image corresponding with a non-uniform distribution of Te within the bulk crystal. Note that non-uniform distribution of impurities around dislocation cores (Cottrell clouds) are known to produce EBIC contrast results (which appear as halos) in other semiconductor materials [7].

Pulse height spectra measurements confirmed that the CZT crystals taken from the middle of the low temperature grown ingots were nuclear detector grade crystals. Figure 7 shows a low temperature grown CZT Frisch collar [8] detection spectra, taken using an [241]Am radiation source, with an FWHM of 2.5 keV (4.2%) at 59.6 keV. The detection spectra clearly show that the low temperature tellurium solvent growth method is capable of producing excellent spectrometer grade CZT radiation detector crystals.

CONCLUSIONS

Detector grade $Cd_{0.9}Zn_{0.1}Te$ single crystals were successfully grown using a low temperature solution growth method, using Te as a solvent. The grown crystals were processed and characterized revealing information detailing defect states and Te inclusions within the bulk of the grown crystals. Thermally stimulated current (TSC) revealed the presence of

Figure 6: (a) IR transmission image (1 x 1 mm²) of CZT crystal (b) EBIC image of a low temperature grown CZT crystal at 80 V reverse bias.

defects/complexes within the bulk of the CZT crystals examined, all of which were attributable to Te inclusions. Electron beam induced current (EBIC) was used on CZT to examine and correlate defect states within the bulk of the grown CZT crystals. Finally, the low temperature grown CZT crystals were fabricated into Frisch collar detectors and their performance was evaluated using pulse height spectra. Future

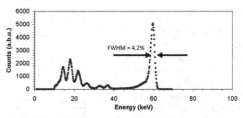

Figure 7: Pulse height spectra for a fabricated Frisch collar detector using a low temperature grown CZT single crystal, using [241]Am source and -700V bias.

efforts will be directed towards further characterization of the bulk CZT material, and improvements to the low temperature crystal growth technique.

ACKNOWLEDGEMENTS

The authors are thankful to Prof. Tangali S. Sudarshan for assistance with TSC characterization.

REFERENCES

1. Egarievwea, S. U., Chena, K. T., Burger, A., James, R. B. & Lissec, C. M., Detection and Electrical Properties of CdZnTe Detectors at Elevated Temperatures. *J. X-ray Sci. and Tech.* **6**, 309-315 (1996).
2. Chattopadhyaya, K., Fetha, S., H. Chena, A. B. & Sub, C.-H., Characterization of semi-insulating CdTe crystals grown by horizontal seeded physical vapor transport. *Journal of Crystal Growth* **191**, 377-385 (1998).
3. Mandal, K. C., Muzykov, P. G., Krishna, R., Hayes, T. & Sudarshan, T. S., Thermally stimulated current and high temperature resistivity measurements of 4H semi-insulating silicon carbide. *Solid State Communications* **151**, 532-535 (2011).
4. Muzykov, P. G., Krishna, R., Das, S., Hayes, T. & Sudarshan, T. S., Characterization of 4H semi-insulating silicon carbide single crystals using electron beam induced current. *Materials Letters* **65**, 911-914 (2011).
5. Krsmanovic, N., Lynn, K. G., Weber, M. H., Tjossem, R. & Gessmann, T., Electrical compensation in CdTe and CdZnTe by intrinsic defects. *Phys. Rev. B* **62**, R16279-R16282 (2000).
6. Soundararajan, R., Lynn, K. G., Awadallah, S., Szeles, C. & Wei, S.-H., Study of defect levels in CdTe using thermoelectric effect spectroscopy. *Journal of Electronic Materials* **25**, 1333-1340 (2006).
7. Maximenko, S., Soloviev, S., Cherednichenko, D. & Sudarshan, T., Electron-beam-induced current observed for dislocations in diffused 4H-SiC P–N diodes. *Appl. Phys. Lett.* **84**, 1576 (2004).
8. Kargar, A., Jones, A. M., McNeil, W. J., Harrison, M. J. & McGregor, D. S., CdZnTe Frisch collar detectors for gamma-ray spectroscopy. *Nuclear Instruments and Methods in Physics Research A* **558**, 497-503 (2006).

Mater. Res. Soc. Symp. Proc. Vol. 1341 © 2011 Materials Research Society
DOI: 10.1557/opl.2011.1272

Improvement of CdMnTe Detector Performance by MnTe Purification

K. H. Kim, A. E. Bolotnikov, G. S. Camarda, R. Tappero, A. Hossain, Y. Cui, G. Yang, R. Gul, and R. B. James
Brookhaven National Laboratory, Upton, NY 11973, USA

ABSTRACT

Residual impurities in manganese (Mn) are a big obstacle to obtaining high- performance CdMnTe (CMT) X-ray and gamma-ray detectors. Generally, the zone-refining method is an effective way to improve the material's purity. In this work, we purified the MnTe compounds combining the zone-refining method with molten Te that has a very high solubility. We confirmed the improved purity of the material by glow-discharge mass spectrometry (GDMS). We also found that CMT crystals from a multiple refined MnTe source, grown by the vertical Bridgman method, yielded better performing detectors.

INTRODUCTION

Although the physical properties of CdMnTe (CMT) and CdZnTe (CZT) are very similar, CMT crystals show poorer performance as X-ray detectors; furthermore, controlling their electrical properties was very difficult due to the high concentrations of residual impurities arising from the starting material [1,2]. Despite these drawbacks, one important difference favoring the usage of CMT rather than CZT is the fact that segregation coefficient of Mn in the CdTe host is close to 1, thereby ensuring a uniform alloy composition within large-volume ingots. Reportedly, cadmium vacancies in CdTe-based semiconductors are successfully compensated by doping with indium, chlorine, or aluminum [1,2].

The concentrations of native points defects (NPDs) in CdTe crystals at room temperatures (RT) is about $10^{13} - 10^{15}$ cm^{-3} [3,4]. Maintaining a concentration of residual impurities below this level ensures their minimal effect on the crystals' electrical properties. Typically, the most effective way to minimize uncontrollable impurities overall by combinations of repeated vacuum-distillation and zone-refining [4,5]. Such methods resulted in impurity contents as low as 10^{15}-10^{16} cm^{-3}, depending upon the particular impurities. In our experiment, we improved the purity of MnTe via the molten Te zone-refining method, and verified it in the better performance of the resulting CMT detector.

EXPERIMENT

The CMT:In crystals were grown by the vertical Bridgman method using a mixture of CdTe (6N), MnTe, and excess Te. At high temperatures, Mn readily reacts with the quartz tubing even if its inner wall is completely carbon coated. The MnTe was synthesized at 1026 C at Te-rich conditions, i.e., with 44% of Mn (4N) and 56% of Te (6N). The source manganese first was cleaned with diluted HNO$_3$ in methanol. Then, vitreous graphite tubing containing Mn and Te was loaded in quartz tube, and sealed under a vacuum of 10^{-6} torr. Then, the temperature was raised at a rate of 0.5 C/min to 960 C to avoid uncontrolled exothermic reactions, and maintained there for one week until the reaction was completed. Finally, we raised the temperature to 1026 C and held it there for 24 hours. We charged the synthesized MnTe in the carbon-coated quartz ampoule with suitable amounts of Te in accordance to the phase diagram and temperature [6]. The heating zone was set to 800 C, and then the ampoule was translated vertically at a speed of 6 mm/day. The effectiveness of zone-refining was verified by glow-discharge mass spectrometry (GDMS).

Four CMT samples of two different sizes, $10 \times 10 \times 2$- and $4 \times 4 \times 11$-mm^3 were prepared from the ingots. The resistivity of those samples was about $2-4 \times 10^{10}$ $\Omega \cdot$cm. The extended defects in samples were inspected via White X-ray Beam Diffraction Topography (WBXDT) [7], an s technique frequently employed to visualize growth- and process-induced defects, such as dislocations, twins, domains, and inclusions. The electrode was prepared by an electro-less method with 5% AuCl$_3$. A standard eV Products test holder with a preamplifier, further shaped and amplified with a standard research amplifier, was employed to measure the ^{241}Am and ^{132}Cs gamma-ray signals, while keeping the temperature around 20 C. The gamma source was uncollimated, and shaping time was set as 2 seconds.

RESULTS AND DISCUSSION

Figure 1 is a cross-sectional view of an as-grown CMT ingot wherein the uniformity of Mn in a CMT slab was measured by micro X-ray fluorescence (μXRF). Usually, the high iconicity of CMT results in low stacking-fault energy, and thus, twins are generated readily. However, controlling growth conditions can prevent the generation of twins. For the first time, we demonstrated homogeneous Mn distribution throughout a whole CMT ingot.

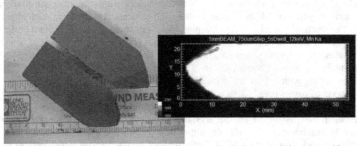

Figure 1. Cross-sectional view of the 1-inch diameter CMT ingot and the Mn uniformity in the CMT slab. Whole area is single grain except first to freeze part. The μ–XRF measurements were acquired at BNL's NSL X-27a beamline, over a 55×22 mm^2 area with a 750- μm step in BNL.

With WBXDT, we can gain information on the type and distribution of structural defects, such as dislocations, inclusions/precipitates, stacking faults, growth-sector boundaries, twins, and low-angle grain boundaries in single crystal materials. The principle of X-ray topography is same as that of Laue diffraction. Passing the white synchrotron beam through a monochromator creates an X-ray topograph when the crystal is set to the Bragg angle for a specific set of lattice planes for the selected X-ray energy. Images from different atomic planes are acquired by orienting the sample to satisfy the Bragg condition for those planes, and orienting the detector to the new scattering angle ($2\theta_B$) to record the image. Depending on operational mode, the system can function in the transmission- or the reflection-mode. The transmission and the reflection mode, respectively, is used for investigating defects in the bulk and surface. Figure 2 shows typical WBXDT images of CMT crystals grown by the vertical Bridgman method. They contains many extended defects, such as low-angle grain boundaries, grain boundaries, and large (20-30 μm) Te inclusions.

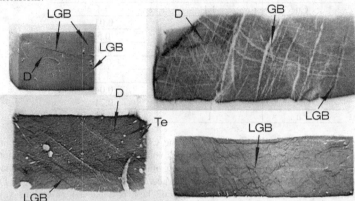

Figure 2. Typical WBXRD data from CMT crystals. They contain extended defects such as grain boundaries (GB), low-angle grain boundaries (LGB), dislocations (D), and Te inclusions (Te).

The segregation coefficient (k) and solubility of elements in molten Te is a crucial factor in purifying raw material by molten Te zone-refining. In CdTe, except for group II elements, most elements have a segregation coefficient of less than 1, and close to 1 for Sn and Mn in CdTe. Table 1 summarizes the Glow-Discharge-Mass-Spectrometer (GDMS) analysis data on the just-synthesized and twice zone-refined MnTe. There are no data available on these kinds of impurities presented in MnTe/CMT material, on the segregation coefficient (k) of impurities, and its effects on the electrical properties. The most abundant impurities present in synthesized MnTe with Mn (4N) and Te (6N) were Cu, Fe, Al, S, Mg, and Si. After purifying the MnTe using a Te solvent, these impurities were greatly reduced as shown in Table 1. However, these impurities levels are still high. In our previous research, we found a photoluminescence (PL) peak located around 0.53 eV just in CMT samples [8]. Maybe one of these impurities is responsible for that PL peak.

Table I. Concentrations of major impurities in MnTe before and after purification.

Elements	Before (ppm wt)	After (ppm wt)
Al	2.1	0.05
Cl	0.19	0.07
Cu	1.1	0.4
Cr	1.3	0.01
Fe	6	0.5
Mg	4.2	0.3
P	0.5	< 0.007
S	40	1.1
Se	0.9	0.04
Si	6	0.21
Zn	1	0.01

The electron mobility-lifetime product of the CMT detectors was calculated from the dependence of peak channel number versus the bias voltage by fitting the Hecht equation; the typical product was $3\text{-}4\times10^{-3}$ cm/V. Figure 3 shows gamma-ray response of the CMT detectors. For the first time, we obtained ^{137}Cs spectra with large-volume CMT detector in virtual Frisch-grid configuration [9]. The energy resolution of 662 keV is about 2.1 % and its value is comparable with that of a good CZT detector.

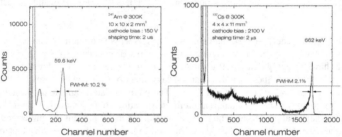

Figure 3. Gamma-ray response of the planar and bar-type virtual Frisch-grid CMT detector on an ^{241}Am and a ^{137}Cs source. The size of the planar and bar-type detector, respectively, is $10\times10\times2$ mm^3 and $4\times4\times11$ mm^3.

CONCLUSIONS

The CMT ingots grown under optimized conditions by the vertical Bridgman method show nearly single-grain crystals and very uniform Mn distribution throughout whole areas. A zone-refining method with molten Te solvent proved an effective way of enhancing material purity and the overall performance of the CMT detectors' performance. The major impurities existing in MnTe, i.e., Cu, Fe, Al, Mg, S, and Si, likely play a key role in electrical properties of

CMT material; our refinement lowered their concentrations, and hence, improved the detectors' performances. More experiments will clarify the segregation tendency of these elemental impurities MnTe- and CMT-materials.

ACKNOWLEDGMENTS

This work was supported by by the U.S. Department of Energy, Office of Nonproliferation Research and Development, NA-22 and the Defense Threat Reduction Agency.

REFERENCES

1. A. Burger, K. Chattopadhyay, H. Chen, J. O. Ndap, X. Ma, S. Trivedi, S. W. Kutcher, R. Chen, and R. D. Rosemeier, *J. Cryst. Growth* **198/199**, 872 (1999).
2. K. Kim, S. Cho, J. Suh, J. Hong, and S. Kim, *IEEE Trans. Nucl. Sci.* **56**, 858 (2009).
3. K.Zanio, Cadmium Telluride, Semiconductors and Semimetals, R.K.Willardson and A.C.Beer (Eds), vol. 13, 1978, New York, San Francisco, London, Academic Press.
4. R.B. James, T.E. Schlesinger, J. Lund, and M. Schieber, "Semiconductors for Room Temperature Nuclear Detector Applications", Semiconductors and Semimetals, Vol. 43, (Academic Press, New York, 1995).
5. W.G. Pfann, "Zone Melting", (New York, Wiley, 1959).
6. C. Reig, V. Munoz, C. Gomez, Ch. Ferrer, and A. Segura, *J. Cryst. Growth* **223**, 349 (2001).
7. G. A. Carini, G. S. Camarda, Z. Zhong, D. P. Siddons, A.E. Bolotnikov, G.W. Wright, B. Barber, C. Arone, and R.B. James, *J. Electro. Mater.* **34**, 804 (2005).
8. K. H. Kim, A. E. Bolotnikov, G. S. Camarda, A. Hossain, R. Gul, G. Yang, Y. Cui, R. B. James, J. Prochazka, J. Franc, and J. Hong, *J. Appl. Phys.* (2011) (in press).
9. A. E. Bolotnikov, G. S. Camarda, G. A. Carini, M. Fiederle, L. Li, D. S. McGregor, W. McNeil, G.W. Wright, and R. B. James, *IEEE Trans. Nucl. Sci.* **53 (2)**, 607 (2006).

Mater. Res. Soc. Symp. Proc. Vol. 1341 © 2011 Materials Research Society
DOI: 10.1557/opl.2011.1107

Characterization of Pd Impurities and Finite-Sized Defects in Detector Grade CdZnTe

M.C. Duff[1] J.P. Bradley[2] Z.R. Dai[2] N. Teslich[2] A. Burger[3] M. Groza[3] and V. Buliga[3]

[1] Savannah River National Laboratory, Aiken, SC 29808, U.S.A.

[2] Lawrence Livermore National Laboratory, Livermore, CA 94550, U.S.A.

[3] Fisk University, Nashville, TN 37208, U.S.A.

ABSTRACT

Synthetic CdZnTe or "CZT" crystals are highly suitable for γ-spectrometers operating at the room temperature. Secondary phases (SP) in CZT are known to inhibit detector performance, particularly when they are present in large numbers or dimensions. These SP may exist as voids or composites of non-cubic phase metallic Te layers with bodies of polycrystalline and amorphous CZT material and voids. Defects associated with crystal twining may also influence detector performance in CZT. Using transmission electron microscopy, we identify two types of defects that are on the nano scale. The first defect consists of 40 nm diameter metallic Pd/Te bodies on the grain boundaries of Te-rich composites. Although the nano-Pd/Te bodies around these composites may be unique to the growth source of this CZT material, noble metal impurities like these may contribute to SP formation in CZT. The second defect type consists of atom-scale grain boundary dislocations. Specifically, these involve inclined "finite-sized" planar defects or interfaces between layers of atoms that are associated with twins. Finite-sized twins may be responsible for the subtle but observable striations that can be seen with optical birefringence imaging and synchrotron X-ray topographic imaging.

INTRODUCTION

Synthetic $Cd_{1-x}Zn_xTe$ or "CZT" crystals are candidate materials for use as a room temperature-based radiation spectrometer. The ternary alloy with 10% Zn, $Cd_{0.9}Zn_{0.1}Te$ has a high band gap of ~1.6 eV and high resistivity form (~10^{10} Ω•cm), which facilitates its use as a simplistic semiconductor spectrometer and imager that requires little power and no cryogenic cooling for operation. Over the last decade, the methods for growing high quality CZT have improved the quality of the produced crystals however various defects remain in these materials that can influence their performance as radiation spectrometers. For example, various structural heterogeneities within the crystals, such as twinning, pipes, grain boundaries (polycrystallinity), elemental impurities and secondary phases (SP) can have an impact on the detector performance [1,2,3,4].

Transmission electron microscopy (HR-TEM) has been used to examine the atom scale defects in SP and the bulk of CZT [1,5,6,7]. The morphology and crystalline plane orientation of the SP that consisted of dendritic and Te-rich material in CZT have been investigated in a highly detailed study using scanning electron microscopy (SEM) and electron backscatter diffraction (EBSD) [8]. We recently characterized SP in a sample of modified vertical Bridgman grown CZT from Yinnel Tech (YT). This material possessed properties that are suitable for its use in γ-radiation detectors. Our study, which used SEM and HR-TEM techniques identified mostly empty 20 μm SP voids in addition to some similarly sized Te-rich SP are published in Duff et al. (2009) [9].

The reason for the formation of the Te-rich SP could not be determined from our study. Elemental impurities in the melt may contribute to their formation. Information on elemental composition of bulk impurities (usually performed by glow discharge mass spectrometry) is often available for CZT materials but information on the composition of such impurities within SP is fairly limited [e.g., ref.10]. The type of growth method used may influence SP formation as well.

Our present study continues with the characterization of this Te-rich material as well as bulk material from the same wafer of material. It discusses the elemental composition and morphology of the impurities that were isolated from the Te-rich SP, which occupied ~10% of the SP in this particular CZT material. It also presents data on the atom scale defects that were observed in the bulk of this material.

EXPERIMENTAL DETAILS

A high performance radiation spectrometer CZT crystal (378) material (received as a 12.1 x 11.3 x 6.8 mm^3 material) was grown according to the Modified Vertical Bridgman (MVB) to have x=0.1 Zn content as in Li et al. procured from YT, South Bend, IN [11]. This type of melt-based growth involves the movement of crucible containing a high purity (typically >7N to 8N) polycrystalline CdZnTe charge through a high temperature furnace. The preparation of this CZT material in this study is described in Duff et al. (2007; 2009) [9,12]. The CZT (as the sub-sampled 378-3 portion) was provided to LLNL for examination by SEM using an FEI Nova 600 Nanolab Dualbeam focused ion beam scanning electron microscope (FIB-SEM) at 5 keV to analyze and produce thin sections of SP-based areas of interest (through ion-beam milling) for HR-TEM studies [as described in ref. 9] as shown in fig. 1.

The HR-TEM studies were performed with a 200 keV FEI Technai20 G2 FEG mono-chromated scanning transmission electron microscope (STEM) with high angle annular dark field (HAADF) detector using a Si(Li) solid state X-ray detector with 0.3 steradians solid angle. For our TEM studies, two thin sections (shown in figs. 2 and 3) from our prior TEM study [9] were used in the current study. However, one of the sections (specifically, the one made from the bottom portion of a SP negative crystal or void, as shown in fig. 2) had to be thinned for our examination of atom scale defects within the surrounding bulk.

Detector performance and methodology information for 378 are also published [12]. Optical birefringence imaging performed with this YT material using 1150 nm IR illumination. The Optical birefringence imaging that we used for our CZT characterization with this and other CZT material are published elsewhere as well [13].

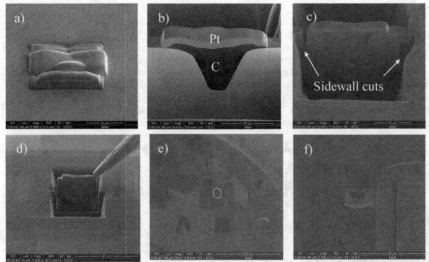

Figure 1. SEM images of the FIB process: a) "pyramid" void depression feature was filled in with carbon (C) using the *in-situ* deposition capabilities of the FIB. The surface of the feature was covered with platinum (Pt). Carbon and Pt layers reduce the effects of potential beam damage during the FIB milling process. In b), the feature after the ion beam has trenched either side of the Pt strap – revealing a cross-sectional sample view. In c), the section is further thinned to ~1 μm and the ion beam is used to make sidewall and under-cuts to enable the extraction of the section. In d), the *in-situ* extraction of the section from the bulk. In e) post attachment of the section to the TEM grid and f) after further ion thinning which produced a ~100 nm section.

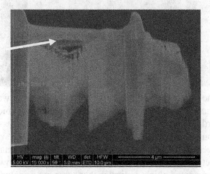

Figure 2. Re-thinning and re-polishing of a second thin section of CZT 378-3 for higher resolution imaging in this study (initial FIB preparation shown in fig. 1). This thin section was isolated in Duff et al. (2009) and was made of bulk single crystal material that was proximal to a void containing nanoparticulate (Cd,Te, Zn, and Si) residue (see arrow near remainder of void).

RESULTS

Characterization of SP impurities

In our prior studies with 378-3, we identified several 20-μm sized hexagonal-shaped entities that were filled with metallic Te (like that shown in fig. 3) on the Te- and Cd-rich faces [9]. In our current study, a closer examination of the composition and crystalline properties of the filled portions of these heterogeneities (SP) revealed the existence of polycrystalline (PC) Te metal, amorphous (AM) CZT and PC CZT. Within the PC Te metal, we observed 40 nm particles that were located at grain boundaries within the Te-rich material as shown in fig. 3a-f and fig. 4a.

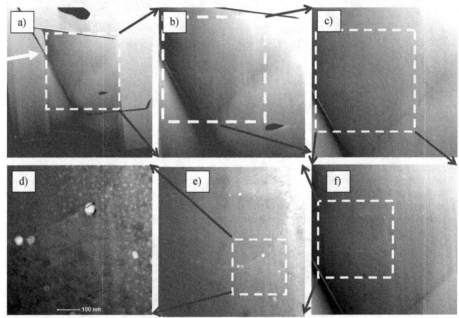

Figure 3. SEM images of the thin section of the Te-rich SP in 378-3 that was later examined using TEM. The teardrop-shaped object (in center right of a) contains a void-like cave. The teardrop shaped object is bordered by polycrystalline (PC) CdZnTe that has an amorphous (AM) CdZnTe rim. The material outside of the AM CdZnTe rim is PC Te. The white arrow in a) points to the boundary between the SP and the bulk single crystal CZT. Images revealing the presence of a nano-size object within the AM Te material in the SP are shown in a) through f). Closer examination of the nano-sized object indicates that some are oriented along a grain boundary inside the metallic PC Te of the Te-rich SP. [Note: Image adapted from fig. 4 in ref. 9 and relative to fig. 4 in ref 9, the images a) and f) are reversed from left to right.]

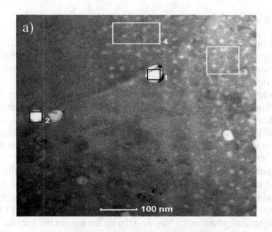

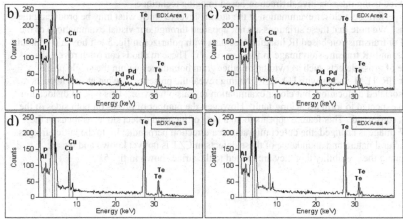

Figure 4. Elemental identification of four various regions in the PC Te metal portion of the Te-rich phase shown in a). The white areas in a) correspond to 40 nm-sized nano particles of PdTe as shown in b) and c). Bulk material analyses by EDX reveal the absence of Pd and a Te-rich material (PC Te metal, according to ref. 9).

These particles were primarily Pd-rich but they were also rich in Te as evidenced by the energy dispersive X-ray (EDX) analyses (shown in fig. 4b-e). Trace levels of Al and P were also found throughout the material regardless of location as shown in fig. 4b-e.

Characterization of line-type defects

The optical birefringence imaging studies with sample 378 revealed little differences in the images that were taken without the cross analyzer as shown in fig. 5 (images in left box). However, several features can be observed in the images taken with the cross analyzer in fig. 5 (see images in box at right).

Several striations appear vertically and at angles throughout the material may indicate the presence of non-uniformity in the electric field. This non-uniformity could potentially impact charge transport and have a detrimental effect on radiation detector performance—although we know that this material is detector grade. Striations have been observed with optical birefringence images in highly resistive detector grade CZT material grown using the traveling heater method by Redlen Technologies (British Columbia, Canada) [13].

We anticipate that these striae represent a certain measurable level of crystalline strain in the material. This level of strain may be induced at the atom level. We have observed subtle striations (or waviness) in CZT using synchrotron X-ray topography imaging studies with material grown from the same boule as well as from YT samples 378 and YT-5 [9,14]. These studies indicate that crystalline strain exists based on the differences in the long range order can be observed using this sensitive diffraction-based imaging technique.

We performed a closer examination at the atomic level to study what may be producing the striations. We note that these striations are not apparent through our visual examinations of the material or transmission-based IR imaging (see data with polarizer in fig. 5 left box and without polarizer; and IR transmission image in fig. 1 in ref. 9). These methods can often reveal the presence of gross defects such as twining or other grain boundaries within the single phase.

Our HR-TEM studies with HAADF imaging reveal the presence of what at first appears to be a dissociated dislocation, but closer examination reveals otherwise. This type of dislocation could be expected to form a stacking fault. However, the number of planes on both sides of the defect is conserved. This feature appears to be a type of inclined defect since, as focus in the HAADF image is changed, the defect migrates in a direction perpendicular to the lattice fringes. The structural nature and abundance of these defects in CZT is not yet known and we are investigating the possibility that they are related to the striae shown in fig. 5.

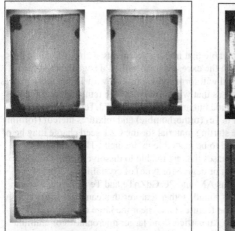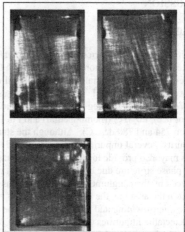

Figure 5. Cross polarized IR images of the same three perpendicular sides of as received YT 378 (12.1 x 11.3 x 6.8 mm^3) material reveal for 1150 nm polarized light with only the polarizer at 45 degrees with vertical (images in box at left) and the polarizer with cross analyzer at 45 degrees (images in box at right).

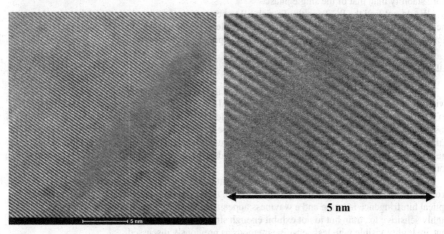

Figure 6. HAADF imaging of atom scale defect in CZT 378-3 from a single crystal area near the void with left) the parent image and right) an inset from a) with 3x magnification.

DISCUSSION

SP impurities

The melt is often Te-rich and it is conceivable that impurity elements such as Pd could be expected to form one or more compounds with the excess Te. Metallic Pd is generally a face centered cubic (fcc) material [15]. Metallic Pd will form solid solutions with Te but these phases are not (primarily fcc) [16]. Examples of phases that will form below the temperature of cubic CdZnTe growth (~1100 deg C for MVB) include but are not limited to: Pd_3Te (body centered cubic or bcc), $Pd_{20}Te_7$ (rhombohedral) and Pd_8Te_3 (orthorhombic) and mixtures thereof (forming between 754 and 780 deg. C). Although the starting material for the CZT seed charge may be of high purity, several impurities can be expected to be insoluble in the melt. These impurity phases may also provide locations for other phases that are unable to dissolve within the solid single phase structure due to their composition or degree (or type) of crystallinity. This is evidenced by the conglomeration of the various AM and PC CdZnTe and Te phases that are present in the area that the nano-phase PdTe is found. Other features that can be observed include micro-twining and grain boundary-type features (such near the large cave as in fig. 3a).

Detectable impurities containing <0.5 μm Au particles and larger microphases of alumina have been observed within SP or in the case of calcium carbonate ($CaCO_3$) within the entirety of the SP in prior studies with MVB-grown CZT material from WSU [10,17]. We have observed nanoparticulate Cd/Zn/Te/Si residue in SP voids in YT 378. It is reasonable to expect that impurities will assimilate in some manner when they are not soluble in the single phase. It is also reasonable that these impurities may reside within another material (like PC Te) that offers more stability than that of the single phase.

We do not find the PC Te in the Te-rich SP to be an artifact of the electron irradiation. Also, the Pd that we observe is not an artifact of the sample preparation because Pd is not used in the electron microscopes that were used for the sample preparation or analyses.

Line-type defects

Marquis et al. (2004, 2007) observed similar "finite-sized" grain boundary defects with {112} twins that formed junctions with {111} twins in gold [as observed in fig. 1 of ref. 18 and fig. 3 of ref. 19]. These examples of offset planes for fcc gold metal are similar to what we have observed with CZT, which is possesses which can be referred to as fcc (i.e., with two offset fcc lattices). Although of interest but not performed in our study, first principles calculations, could yield information about the total energy of the boundary per unit length and level of strain that is compensated for in this material (in ref. 18).

On a larger scale, the inclined planar defects that are observed at the atom scale may propagate enough lattice strain to be visible through techniques (e.g., as striae as viewed with optical birefringence imaging and a waviness appearance in X-ray topography) and that are highly sensitive to strain but to not exhibit enough strain to cause large scale grain boundaries that are highly visible with less sensitive methods as previously discussed.

We noted that our study findings differ from those of Zeng et al. (2009), who examined abrupt, large-scale twin boundaries and sub-boundary networks with bulk CZT grown using the MVB method with HR-TEM. Our observations examined differences on the atom scale rather than that on the larger 200 to 1000 nm scale.

CONCLUSIONS

This study presents new observations about the morphology and composition of the nano-sized defects in detector grade CZT. We find although Pd is used as a coating for many electron microscopy analyses (in other electron microscopes), the Te-rich Pd nano particles that we observe are not the results of Pd that has been used for this purpose.

ACKNOWLEDGMENTS

Work supported by US DOE - National Nuclear Security Administration, through the Office of Nonproliferation and Verification Research and Development (NA-22) and National Science Foundation through the Fisk University Center for Physics and Chemistry of Materials (CPCoM), Cooperative Agreement CA: HRD-0420516 (CREST program) and through and from US DOE NA-22 Grant No. DE-FG52-05NA27035. We thank Dr. Doug Medlin (of LLNL) for his helpful comments.

REFERENCES

1. J. R. Heffelfinger, D. L. Medlin, and R. B. James. *MRS Symp. Ser.* **487**, 49-54 (1998).
2. M. Schieber, T. E. Schlesinger, R. B. James, H. Hermon, H. Yoon, and M. Goorsky. *J. Cryst. Growth* **237-239**, 2082 (2002).
3. C. Szeles and M. C. Driver. *SPIE Proc.* **3446**, 1 (1998).
4. J. Shen, D. K. Aidun, L. Regel, and W. R. Wilcox. *Cryst. Growth* **132**, 250-260 (1993).
5. T. Wang, W. Jie, and D. Zeng. *Mater. Sci. Engin. A* **472**, 227-230 (2008).
6. S. Rai, S. Mahajan, S. McDevitt, and C. J. Johnson. *J. Vac. Sci. Tech.* **B9**, 1892 (1996).
7. D. Zeng, W. Jie, T. Wang, and H. Zhou. *J. Crystal Growth* **311**, 4414-4417 (2009).
8. C. Henager, D. J. Edwards, A. L. Schemer-Kohrn, M. Bliss, and J. E. Jaffe. *J. Crystal Growth* **312**, 507–513 (2010).
9. M. C. Duff, D. B. Hunter, A. Burger, M. Groza, V. Buliga, J. P. Bradley, G. Graham, Z. R. Dai, N. Teslich, D. R. Black, and A. Lanzirotti. *J. Mater. Res.* **24**, 1361-1367 (2009).
10. M. C. Duff, K. G. Lynn, K. Jones, Z. R. Dai, J. P. Bradley, and N. Teslich. *SPIE Proc.* **7449** 74490N (2009) (oral and written publications).
11. L. Li, F. Lu, C. Lee, G. Wright, D. R. Rhiger, S. Sen, K. S. Shah, M. R. Squillante, L. Cirinano, R. B. James, A. Burger, P. Luke, and R. Olson. *SPIE Proc.* **4784**, 76 (2003).
12. M. C. Duff, D. B. Hunter, P. Nuessle, D. R. Black, H. Burdette, J. Woicik, A. Burger, and M. Groza. *J. Elect. Mater.* **36**, 1092-1097 (2007).
13. S. A. Awadalla, J. Mackenzie, H. Chen, B. Redden, G. Bindley, M. C. Duff, A. Burger, M. Groza, V. Buliga, J. P. Bradley, Z. R. Dai, N. Teslich, and D. R. Black. *J. Crystal Growth* **312**, 507–513 (2010).
14. M. C. Duff, D. B. Hunter, A. Burger, M. Groza, V. Buliga, and D. R. Black. *Appl. Surf. Sci.* **254**, 2889-2892 (2008).
15. M. Harada, K. Asakura, Y. Ueki, and N. Toshima. *J. Phys. Chem.* **96**, 9730-9738. (1992).
16. H. Okamoto. *J. Phase Equilibria* **13**, 73-79 (1992).
17. M. C. Duff, K. G. Lynn, K. Jones, R. Soundararajan, J. P. Bradley, H. Ishii, J. Aguiar, and P. Wozniakiewicz. *SPIE Proc.* **7805**, 74490N (2010).
18. E. A. Marquis, J. C. Hamilton, D. L. Medlin, and F. Leonard. *Phys. Rev. Lett.* **93**, 1-4 (2004).
19. E. A. Marquis, D. L. Medlin, and F. Leonard. *Acta Materialia* **55**, 5917-5923 (2007).

Mater. Res. Soc. Symp. Proc. Vol. 1341 © 2011 Materials Research Society
DOI: 10.1557/opl.2011.1274

Effects of dislocations and sub-grain boundaries on X-ray response maps of CdZnTe radiation detectors

A. Hossain[1], A. E. Bolotnikov[1], G. S. Camarda[1], Y. Cui[1], R. Gul[1], K. Kim[1],
B. Raghothamachar[2], G. Yang[1] and R. B. James[1]

[1]Brookhaven National Laboratory, Upton, NY, USA
[2]Stony Brook University, Stony Brook, NY, USA

ABSTRACT

The imperfect quality of CdZnTe (CZT) crystals for radiation detectors seriously diminishes their suitability for different applications. Dislocations and other dislocation-related defects, such as sub-grain boundaries and dislocation fields around Te inclusions, engender significant charge losses and, consequently, cause fluctuations in the detector's output signals, thereby hindering their spectroscopic responses. In this paper, we discuss our results from characterizing CZT material by using a high-spatial-resolution X-ray response mapping system at BNL's National Synchrotron Light Source. In this paper, we emphasize the roles of these dislocation-related defects and their contributions in degrading the detector's performance. Specifically, we compare the effects of the sub-grain- and coherent twin-boundaries on the X-ray response maps.

KEYWORDS: CdZnTe, radiation detectors, dislocations, and crystal defects

INTRODUCTION

The inadequate quality of today's CdZnTe (CZT) crystals is the main factor limiting their performance and availability of CZT radiation detectors. Previously, we demonstrated that dislocations and other dislocation-related defects, such as sub-grain boundaries and the dislocation fields around Te inclusions, cause significant charge losses and, consequently, fluctuations in the output signals that obstruct the spectroscopic responses of CZT detectors [1]. In this paper, we present our new experimental results from characterizing CZT material using a high-spatial resolution X-ray mapping system at BNL's National Synchrotron Light Source. The results emphasize the roles of impurities and secondary phases accumulated by dislocation-related defects in causing such degradation. Specifically, we compare the effects of the sub-grain- and coherent twin-boundaries on the X-ray response maps.

Sub-grain boundaries are an important class of defects, in addition to impurities and Te inclusions, which occur in all commercial CZT materials, regardless of the growth techniques or their vendors. Usually, the vendors cannot specify the content of sub-grain boundaries in their crystals when they sell them, because they mainly employ IR screening methods that are inadequate for identifying sub-grain boundaries. The most effective techniques, unavailable to the vendors for routine material screening, are white beam X-ray diffraction topography. Chemical etching of crystal surfaces, e.g., Saucedo solution [2], is also helpful to trace out such defects. A large variety of sub-grain boundaries is classifiable by their effects on the device's performances: Some likely are of little harm, and, in many cases, can be neglected entirely. Our previous measurements indicate that the effects of the boundaries are primarily related to dislocations that accumulate impurities, secondary phases, and Te inclusions, and to space charges that, in turn, trap carriers and alter the local distributions of the electric field. If this statement is correct, then coherent twin boundaries, a special case of boundaries without dislocations, should not affect charge transport. The goal of our measurements was to verify these predictions.

EXPERIMENTAL

We tested several CZT crystals of different dimensions containing twins but herein, we describe the results obtained from detailed measurements on only one of them, a 6x6x15 mm^3 parallelepiped-shaped sample, that are typical of what we observed in the other crystals. To reveal the locations of the twin- and sub-grain- boundaries, we examined the crystals using white beam X-ray diffraction topography and IR transmission microscopy. Then, we mapped the X-ray response from all side surfaces of the crystals by attaching contacts on these corresponding side surfaces. We used a highly collimated X-ray beam at BNL's National Synchrotron Light Source to measure the X-ray response maps for all six side-surfaces oriented perpendicularly to the beam. For each location of the beam, we measured the pulse-height spectra, evaluated their photo-peaks positions, and then plotted them as photopeak-position maps. Each of these techniques was detailed in our previous publications [3-6].

RESULTS AND DISCUSSION

Twins are a good example of boundaries without dislocations. Dislocations often are not visible under optical- or IR-microscopes; they are apparent only when decorated by Te inclusions and secondary phases, as we reported earlier [7]. We tested several CZT crystals of different dimensions, each containing twins. As mentioned, we show representative results obtained from a 6x6x15 mm^3 parallelepiped-shaped sample. Fig. 1(a) shows optical photographs of this detector whose surfaces first were etched chemically with the Saucedo solution to reveal the location of the twins inside the crystal. Fig. 1(b) depicts the diffraction topographic images of four large surfaces of this crystal, measured by employing a white X-ray beam. We note that straight planes represent the twin boundaries in the images and in contrast, the diffraction topographs show the features of a curved twin caused by internal stresses in the crystal, as well as the sub-grain boundaries crossing the twin.

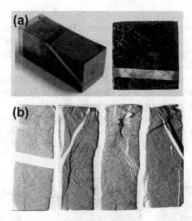

Figure 1. (a) Photographs of a 6x6x15 mm³ detector used for our measurements whose surfaces were etched chemically to reveal the twins exiting the crystal's side surfaces. (b) White beam X-ray diffraction topographs of the detector's four side surfaces, revealing twin- and sub-grain boundaries exiting them.

We used IR transmission microscopy to verify that twin boundaries contain no Te inclusions that otherwise would indicate the presence of dislocations. Fig. 2 is a volumetric image of the 1.1x1.5x6 mm³ parallelepiped region extending from one side of the crystal to another, and crossing the twin. To generate this image of the Te inclusions, we acquired a stack of IR planar images across the sample thickness (6 mm) with 20-μm steps. Then, with these images, we reconstructed a 3D view of the inclusions. Using IR microscopy, we confirmed that twin boundaries contain no Te inclusions (no dislocations) except for the regions where sub-grain boundaries cross the twin's coherent boundaries (free from dislocations).

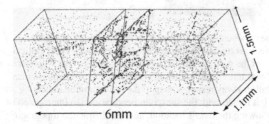

Figure 2. A volumetric IR image of the 1.1x1.5x6-mm³ parallelepiped region extending from one side of the crystal to the other, and crossing the twin. To generate this image of the Te inclusions, we took a stack of planar images in depth across the crystal thickness with 20-μm steps.

We carried out X-ray response mapping (resolution 25- 50 μm²) from all six surfaces after attaching contacts on the corresponding side surfaces. These measurements confirmed that the dislocation-free coherent twin boundaries do not adversely affect charge transport. This supports our previous findings that the detrimental effects of the sub-grain boundaries are attributable entirely to impurities, secondary phases, and Te inclusions that might accumulate inside the sub-grain boundaries [1]. The charge-carrier response map of three representative surfaces of the detector (1, 2, & 3) against the well-displayed twin further validates our earlier findings, as illustrated in Fig. 3. The contrast features in the response map do not correlate with the twin boundary. On the other hand, the correlations between impurities and secondary phases accumulated inside the sub-grain boundaries clearly are visible as dark areas (i.e., regions with higher trapping) in the X-ray response maps.

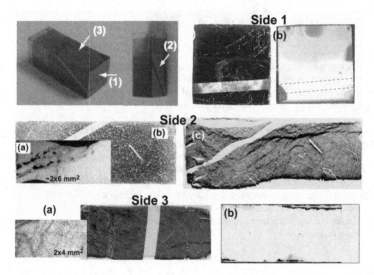

Figure 3. The images of the representative CZT detector crystal, X-ray diffraction images, and response maps were obtained from marked surfaces. **(Side 1)** (a) Optical image showing the location of the twin exiting the top surface; and, (b) the whole-area X-ray response map; the twin was not visible in the high resolution X-ray response map. **(Side 2)** (a) The high-resolution X-ray response map of a ~2x6 mm² area indicating no correlations with the twin, but rather, with the sub-grain boundaries; (b) Photograph of the etched crystal surface showing the twin's location; and, (c) diffraction topography of the crystal's side surface 1 showing the location of the twin and a network of sub-grain boundaries correlated with the dark features in the high-resolution X-ray response map in (a). **(Side 3)** (a) The high-resolution X-ray response map of a 2x4 mm² area showing the correlations with the sub-grain boundaries visible in the diffraction topograph of the crystal's corresponding surface; and (b) whole-area X-ray response map displaying no correlation with the twin.

64

CONCLUSIONS

We found no correlations between the X-ray response maps and the coherent twin boundaries, validating our belief that dislocation-free coherent twin boundaries have no effect on charge transport in CZT detectors. This supports our previous findings that the effects of sub-grain boundaries are attributable entirely to impurities, secondary phases, and Te inclusions that accumulate inside the boundaries. A straightforward way to minimize the effects of the sub-grain boundaries (and dislocations) is to maintain the lowest possible concentrations of impurities and Te inclusions in the crystal.

ACKNOWLEDGEMENT

This work was supported by U.S. Department of Energy, Office of Nonproliferation Research and Development, NA-22, and the Defense Threat Reduction Agency (DTRA).

REFERENCES

1. A. E. Bolotnikov, S. Babalola, G. S. Camarda, Y. Cui, R. Gul, S. U. Egarievwe, P. M. Fochuk, M. Fuerstnau, A. Hossain, F. Jones, K. H. Kim, O. V. Kopach, L. Marchini, B. Raghothamachar, R. Taggart, G. Yang, L. Xu, and R. B. James, "Correlations between crystal defects and performance of CdZnTe detectors", accepted for IEEE Trans. Nucl. Sci., NS, 2011.
2. E. Saucedo, P. Rudolph and E. Dieguez, *J. Cryst. Growth* **310**, p. 2067, 2008.
3. G. A. Carini, A. E. Bolotnikov, G. S. Camarda, G. W. Wright, L. Li, and R. B. James, "Effect of Te inclusions on the performance of CdZnTe detectors", Appl. Phys. Lett. 88, p.143515, 2006.
4. G. S. Camarda, A. E. Bolotnikov, G. A. Carini, and R. B. James, "Effects of Tellurium inclusions on charge collection in CZT Nuclear Radiation Detectors", in Countering Nuclear and Radiological Terrorism, edited by S. Aprkyan and D. Diamond, Springer, 2006, pp. 199-207.
5. G. S. Camarda, N. M. Abdul-Jabbar, S. Babalola, A. E. Bolotnikov, Y. Cui, A. Hossain, E. Jackson, H. Jackson, J. R. James, A. L. Luryi, M. Groza, A. Burger, and R. B. James, "Characterization and Measurements of CZT Material: Novel Techniques and Results", in Proceedings of SPIE Hard X-Ray and Gamma-Ray Detector Physics IX, Vol. 6706, edited by R. B. James, A. Burger and L. A. Franks (SPIE, Bellingham, WA, 2007), 670605.
6. A. E. Bolotnikov, G. S. Camarda, Y. Cui, A. Hossain, G. Yang, H. W. Yao, and R. B. James, "Internal electric-field-lines distribution in CdZnTe detectors measured using X-ray mapping", IEEE Trans. Nucl. Sci., NS 56, n. 3, pp. 791-794, 2009.
7. A. Hossain, A. E. Bolotnikov, G. S. Camarda, Y. Cui, G. Yang, and R. B. James, Defects in cadmium zinc telluride crystals revealed by etch-pit distributions, *Journal of Crystal Growth* **310** (2008) 4493–4498.

Mater. Res. Soc. Symp. Proc. Vol. 1341 © 2011 Materials Research Society
DOI: 10.1557/opl.2011.1206

Study of Structural Defects in CdZnTe Crystals by High Resolution Electron Microscopy

A. Hossain, A. E. Bolotnikov, G. S. Camarda, Y. Cui, R. Gul, K-H. Kim, K. Kisslinger, D. Su,
G. Yang, L. H. Zhang, and R. B. James
Brookhaven National Laboratory, Upton, NY 11973, USA

ABSTRACT

We investigated defects in CdZnTe crystals produced from various conditions and their impact on fabricated devices. In this study, we employed transmission and scanning transmission electron microscope (TEM and STEM), because defects at the nano-scale are not observed readily under an optical or infrared microscope, or by most other techniques. Our approach revealed several types of defects in the crystals, such as low-angle boundaries, dislocations and precipitates, which likely are major causes in degrading the electrical properties of CdZnTe devices, and eventually limiting their performance.

INTRODUCTION

Today's CdZnTe (CZT) radiation detectors suffer from several problems related to material uniformity and hampered charge transport that degrade the devices' overall performance [1-4], and hence restrict their widespread deployment for applications in national security and medical imaging. Such deterioration in the detector's performance results from various point- and extended defects, like vacancies, impurities, dislocations, and precipitates. Unstable conditions during crystal growth and post-growth treatment are likely responsible for these defects. To characterize them, and better understand the challenges they pose for detectors, conventional techniques, such as visualizing dislocation-etched pits, IR transmission microscopy, and white X-ray diffraction topography have been used; however, they provide information on bulk properties only. Therefore, as new technologies become available there is an opportunity and a need to use them to observe directly the nano-structural and other related defects in CZT to clarify in-depth the origins of the formation of extended and point defects, and the crystals' compositional distribution. Electron microscopy offers a variety of techniques that give the ability to characterize the microstructure and the chemical compositional distribution of materials down to the nanoscale. High-resolution electron microscopy (HREM) can afford information about the crystal's lattice and its atomic structure that is far more detailed than is currently known; such comprehensive data is essential to identifying the sources of the defects. We have been systematically characterized such structural defects in CZT crystals grown by different methods. Our objective is to identify and to quantify various types of defects, their concentrations, and origins. In this study, we detailed the structural features and local chemical composition of CZT crystals grown via different growth methods, using high-resolution transmission electron microscopes (TEM/STEM).

Our data on the nano-scale demonstrated distinct structural defects in the crystals, such as interstitial atoms, low-angle boundaries, dislocations, and precipitations that likely are major causes in degrading the electrical properties of the devices, and eventually limiting their performance.

EXPERIMENTAL DETAILS

We studied three CZT samples grown by the high-pressure Bridgman method by three different vendors. We selected the small pieces of CZT samples from a cleaved plane of each ingot, lifted off a tiny portion using a special TEM sampling tool, and Pt soldered it on to a 3-mm-diameter copper holder. The samples were ground down to a few tens of nanometer, so that it roughly approximated the mean-free-path of electrons transmitted through the sample using a focused ion beam (FIB). The CZT TEM analysis was undertaken using a HRTEM (JEOL-2100F, Hitachi HD2700C) with an electron energy of 300 kV. We employed the SEM-EDS (JOEL7600F) for characterizing the chemical composition of the local area in the CZT TEM sample.

RESULTS AND DISCUSSION

Selected Area Diffraction

Fig. 1 displays typical representative images of a selected area diffraction (SAD) pattern on a CZT [111] projection from an area chosen randomly, in which we have labeled the planes' indexes. The periodicity and symmetry of this diffraction pattern (DP) pattern is consistent with the standard zinc-blende diffraction pattern on the [111] projection. Using similar images we observed irregular diffraction features (circled) in this particular SAD pattern, which indicates the presence of foreign atoms lying in the lattice plane. Much effort was made to locate them in the specimens. Using our high-resolution electron microscope, we randomly captured and surveyed images of the lattice planes and traced some foreign objects, like the one shown in fig. 1(b). However, we have not yet identified those objects and their concentrations, nor defined their impact on the detector's performance.

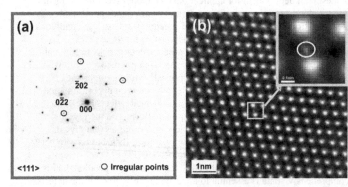

Figure 1. TEM micrograph of CZT crystal (a) Selected area diffraction pattern (inverted color) of crystal structure at the <111> plane, wherein some irregular points indicate the existence of foreign atoms; (b) Interstitial atom was traced in the lattice plane.

Dislocations

We randomly captured the images of the lattice planes from all three specimens; all contained various defects distributed uniformly throughout the samples, including stacking faults and dislocations arranged in linear configurations or in loops. Fig. 2 demonstrates the nanostructure of dislocations that we repeatedly observed in CZT. Fig. 2(a) displays a lattice plane with two lines of edge dislocations wherein two extra half-planes of atoms are introduced midway through the crystal, distorting the nearby planes of atoms; this results in an increment in the system energy of the surroundings. The planes move in response to shear stress, applied from one side of the crystal structure that passes through the planes of atoms, breaking and joining the bonds within them until it reaches the grain boundary. The planes have the tendency to move together to reduce the system's energy. Stress also is generated inside the crystal during the cooling process after growth, and, consequently, parts of the lattices glide along each other to decrease the system's energy.

The TEM micrograph in Fig. 2(b) shows stacking faults, viz., another type of structural defect repeatedly observed in the specimens. Such defects are generated by one- or two-layer interruptions in the stacking sequence of the crystal structure. These disruptions carry a certain energy that determines the width of the stacking fault. The potential energy of the inner layers of the faults has more than the outer side that drives the stacking faults to disappear, turning into dislocation loops. We found high concentrations of these defects in all three samples compared to other defects, such as impurities and vacancies.

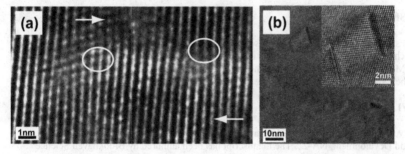

Figure 2. TEM images of CZT crystals (a) Two edge dislocations apparent in a lattice plane; the direction of movement due to stress is shown by the arrows; (b) Stacking fault turning into a dislocation loop.

Precipitates

As we demonstrated previously, Te inclusions play a critical role in the performance of devices [5-7]. Their sizes range from one to several hundred μm, and thus, they are easily seen by scanning electron microscopy and infrared microscopy. However, Te precipitates are usually too small (in the range of nanometers) to trace by conventional methods, and thus, their influence on the devices' performance is not well known. We analyzed our TEM micrographs shown in Fig. 3. We believe that the small, dark, irregularly shaped structures distributed randomly in the matrix are Te precipitates. Although we have not yet verified our claim by compositional analysis, the size of these objects and the Z contrast of the micrographic image, as well as evidence consistent with other published reports [8-9], support our assertion.

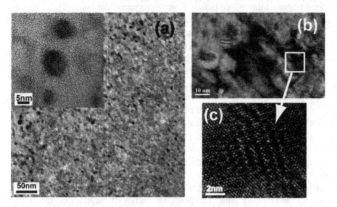

Figure 3. (a) A low-magnification TEM micrograph; the small dark spots are Te precipitations distributed randomly in the matrix; (b) Irregularly shaped black spots with Fresnel fringes, i.e., evidence of Te precipitations; and (c) Magnified image of Te precipitates.

CONCLUSIONS

These studies demonstrated on a nano-scale the presence of several types of defects in CZT crystals, such as stacking faults, dislocations and precipitations distributed throughout the CZT materials. More data are needed to understand the nature of these defects and to quantify the impact of their size and concentrations on the devices' performance.

ACKNOWLEDGEMENT

This work was supported by U.S. Department of Energy, Office of Nonproliferation Research and Development, NA-22. The manuscript has been authored by Brookhaven Science Associates, LLC under Contract No. DE-AC02-98CH1-886 with the U.S. Department of Energy. The United States Government retains, and the publisher, by accepting the article for publication, acknowledges, a world-wide license to publish or reproduce the published form of this manuscript, or allow others to do so, for the United States Government purposes.

REFERENCES

1. C. Szeles and E. E. Eissler, Semiconductors for Room Temperature Radiation Detector Applications II, Vol. 487, edited by R. B. James, T. E. Schlesinger, P. Siffert, W. Dusi, M. R. Squillante, M. O'Connell and M. Cuzin (Materials Research Society, Pittsburgh, PA, 3 (1998).
2. J. R. Heffelfinger, D. L. Medlin, and R. B. James, Semiconductors for Room Temperature Radiation Detector Applications II, Vol. 487, edited by R. B. James, T. E. Schlesinger, P.

Siffert, W. Dusi, M. R. Squillante, M. O'Connell and M. Cuzin (Materials Research Society, Pittsburgh, PA, 33(1998).

3. T. E. Schlesinger, J. E. Toney, H. Yoon, E. Y. Lee, B. A. Brunett, L. Franks, and R. B. James, *Mat. Sci. Eng.* R32, 103 (2001).

4. P. Rudolph, *Cryst. Res. Technol.* **38**, no. 7-8, 542 (2003).

5. G. A. Carini, A. E. Bolotnikov, G. S. Camarda, G. W. Wright, L. Li, and R. B. James, *Appl. Phys. Lett.* **88**, 143515 (2006).

6. A. E. Bolotnikov, G. S. Camarda, G. A. Carini, Y. Cui, L. Li, and R. B. James, *Nucl. Instr. Meth.* **A571**, 687 (2007).

7. A. E. Bolotnikov, G. S. Camarda, G. A. Carini, Y. Cui, K. T. Kohman, L. Li, M. B. Salomon, and R. B. James, *IEEE Trans. Nucl. Sci.,* **NS-54**, no. 4, 821(2007).

8. G. Li, S-J. Shih, Y. Huang, T. Wang, and W. Jie, *Journal of Crystal Growth* **311**, 85 (2008).

9. Y.Y. Loginov, P. D. Brown, K. Durose, Structural defect formation in II-VI semiconductors, M. Logos, ISBN 5-94010-214-X, (2003).

Modelling and Calculations

Mater. Res. Soc. Symp. Proc. Vol. 1341 © 2011 Materials Research Society
DOI: 10.1557/opl.2011.1507

Performance evaluation of neutron detectors incorporating intrinsic Gd using a GEANT4 modeling approach

Abigail A. Bickley[1], Christopher Young[1], Benjamin Thomas[1], John W. McClory[1], Peter A. Dowben[2], James C. Petrosky[1]

[1]Department of Engineering Physics, Air Force Institute of Technology, 2950 Hobson Way, Wright-Patterson Air Force Base, OH 45433-7765, U.S.A.
[2]Department of Physics and Astronomy, University of Nebraska - Lincoln, 855 North 16th St, □ Lincoln, NE 68588-0299, U.S.A.

ABSTRACT

Solid-state neutron detectors from heterostructures that incorporate Gd intrinsically or as a dopant may significantly benefit from the high thermal neutron capture cross section of gadolinium. Semiconducting devices with Gd atoms can act as a neutron capture medium and simultaneously detect the electronic signal that characterizes the interaction. Neutron capture in natural isotopic abundance gadolinium predominantly occurs via the formation of ^{158m}Gd, which decays to the ground state through the emission of high-energy gamma rays and an internal conversion electron. Detection of the internal conversion electron and/or the subsequent Auger electron emission provides a distinct and identifiable signature that neutron capture has occurred. Ensuring that the medium responds to these emissions is imperative to maximizing the efficiency and separating out other interactions from the radiation environment. A GEANT4 model, which includes incorporation of the nuclear structure of Gd, has been constructed to simulate the expected device behavior. This model allows the energy deposited from the decay of the meta-stable state to be localized and transported, providing for analysis of various device parameters. Device fabrication has been completed for Gd doped HfO_2 on n-type silicon, Gd_2O_3 on p-type silicon and Gd_2O_3 on SiC for validation of the code. A preliminary evaluation of neutron detection capabilities of these devices using a GEANT4 modeling approach is presented.

INTRODUCTION

Strong motivation exists for developing new technologies capable of detecting neutron sources. The detection of neutrons has received renewed interest due to concerns about the proliferation of special nuclear materials (SNM) and global terrorism. Of particular interest is the detection of isotopes of uranium and plutonium. Plutonium has a measurable neutron emission rate owing to its unique spontaneous fission rate, while uranium neutron emission can be enhanced using active interrogation methods. However, neutron emission rates are low, thus detectors possessing a large neutron capture cross section and high efficiency are required. Furthermore, the neutron energy spectra of SNM peaks in the MeV region while the neutron capture cross sections of traditional detection media are strongly biased towards thermal energies. An example of this is provided by the ^{3}He proportional counter, which can be used to detect both thermal and fast neutrons via the $^3_2He + ^1_0n \rightarrow ^3_1H + ^1_1p$ reaction. Recent shortages in the global supply of ^{3}He have led to rationing of this isotope [1] and reinvigorated research in alternative technologies.

The neutron capture properties of gadolinium provide an attractive alternative for detecting a low flux of neutrons. Natural gadolinium's absorption cross section for thermal neutrons is 46,000 barns, while the 15.65% abundant ^{157}Gd has a cross section of 255,000 barns. In addition, the cross section remains higher than that of ^{3}He, ^{6}Li and ^{10}B neutron detection reactions up to ~400 meV, and thus reduces the volume of moderator required for efficient detection [4].

Advances in solid-state device fabrication techniques now allows for inclusion of gadolinium in semiconductor devices, both intrinsically and as a dopant [4-6]. This permits the material to act simultaneously as a neutron conversion and electronic medium. Neutron capture in gadolinium predominantly occurs via the formation of ^{158m}Gd, which decays to the ground state through the emission of gamma rays and an internal conversion electron due its high internal conversion coefficient. Detection of the internal conversion electron and/or the subsequent Auger electron emission can provide a distinct and identifiable electronic signature of neutron capture, since these processes occur at well-known discrete energies. Greater than 96% of conversion electrons from neutron capture on ^{157}Gd and the subsequent decay of ^{158m}Gd have energies of 79 and 182 keV. Auger electrons from K and L shell transitions have energies in the 30-50 keV and 1-10 keV ranges, respectively. These energy ranges present two competing challenges. First, the total resulting charge pulses are small and detection requires a semiconducting device with a low leakage current coupled with a low noise readout system. Second, the mean free path of the secondary radiation is large relative to typical semiconductor device material thicknesses. To produce a recognizable trigger condition, the electron must deposit its full energy in the device and the energy must be collected. As a point of reference, a 79 keV electron deposits nearly all of its energy in ~40 μm of silicon and in ~25 μm of gadolinium [7], whereas the depletion regions of the devices of interest typically have widths of <10 μm [8].

To determine whether a solid-state neutron detector constructed from heterostructures incorporating Gd can be used to efficiently detect neutrons, a dual pronged approach including both Monte-Carlo simulation and experimental testing was used. Three different types of devices were constructed: 1) Gd doped HfO$_2$ on n-type silicon [6], 2) Gd$_2$O$_3$ on p-type silicon [9] and 3) Gd$_2$O$_3$ on SiC [8]. The experimental response of these diodes to a neutron flux is reported elsewhere [6,8,9]. The focus of this paper is a GEANT4 [10] model that has been constructed based on the physical parameters of these diodes to simulate the expected device behavior. The model includes a full incorporation of the nuclear structure of Gd, thus allowing the energy deposited from the decay of ^{158m}Gd to be localized and propagated. Of interest from the model are the fraction of energy deposited by each electron and the total energy per neutron capture event deposited in the active volume of the detector.

GEANT4 MODIFICATIONS

The GEANT4 toolkit is a Monte Carlo based software package designed to simulate the transport of particles through matter [10] and can be modified or supplemented as needed by the end user. This feature is critical in the present situation to fully model the nuclear structure of gadolinium. GEANT4 is distributed with independent data libraries for some physics processes. Of interest is the Photon Evaporation data for ^{158}Gd, which is compiled from the Evaluated Nuclear Structure Data File (ENSDF) and maintained by the National Nuclear Data Center, Brookhaven National Laboratory [11]. The most recent version of the Photon Evaporation library available is version 2.1 [12]. Within the ^{158}Gd data file the maximum nuclear energy level included is 3921 keV. Ali, *et al.* have reported additional excited state levels [13] up to the neutron separation energy, 7937.39±0.12 keV [14], that contribute significantly to the first two excited states of ^{158}Gd and are significant sources of internal conversion electrons. These excited states must be included in the GEANT4 model to properly simulate the expected decay spectrum. Additional information regarding prompt gamma rays from neutron activation of ^{157}Gd is available from the International Atomic Energy Agency (IAEA) [15].

Modifications were made to the GEANT4 photon evaporation data file for ^{158}Gd to include the thirteen most significant gamma transitions not previously simulated (**Table I**). Determination of which energies to include was based on the relative intensity of the transition provided in Table 1 of [13]. Implementing nuclear levels and gamma transition energies requires knowledge of the level energy, gamma energy, transition probability, transition type and internal conversion coefficients [16]. The internal conversion coefficients were determined using a fit to the data provided in [17] in the energy range of interest for this research. A sizeable number of allowed transitions tabulated in [15] have not yet been added to GEANT4 because they were either deemed to have a statistically insignificant contribution to the production of internal conversion electrons or insufficient information was available in the literature. Gamma energies for these transitions predominantly fall within 6-8MeV. This results in a non-physical dip in the gamma continuum at high energies in the modified GEANT4 simulation relative to the experimental data, **Figure 1** (left). A comparison of the original and modified simulated gamma spectra reveals a significantly improved distribution with respect to the available experimental data after modifications were applied. This comparison would benefit from additional experimental data for low energy gammas (<1MeV) where detection efficiencies tend to be low.

Table I: ^{158}Gd nuclear energy levels and transitions added to the photon evaporation library.

E Level (keV)	γ Energy (keV)	Transition [11]	Transition [15]	Comment
7937.4	6750.3	E2	E2	
	6420.3	E2	E2	
	5972.3		E1	
	5678.0	E2	E1	Added as E2
7857.9	6750.2	(E2+M1)	E2	Added as E2+M1
	6672.3	(E2+M1)	E2	Added as E2+M1
	5676.8		E1	
	5177.8		E1	
7676.0	6757.9	E1		
	6750.0	E2	E2	
	5659.8		E2	
	4736.1		E1	
	2939.9		E1	

SIMULATED EXPERIMENT

To determine the expected response of a gadolinium based diode in a thermal neutron flux, a physical model of the device was constructed in GEANT4. The characteristics were selected to be representative of the fabricated diodes. A layer of 250 nm thick HfO$_2$ doped with 15% natural gadolinium was deposited on a 1.25 mm thick layer of silicon. The active volume of the diode capable of collecting charge was designated to be the entire gadolinium layer and a portion of the silicon layer up to a depth of 3μm to account for diffusion of ionization induced charge. This configuration is representative of the diode reported in [6] operating at a reverse bias voltage of 5V. Neutrons with an energy of 0.03 eV were perpendicularly incident on the gadolinium layer of the diode. A detailed list of the physics processes simulated for each major particle species is provided in **Table II**. A total of 10^6 neutrons were simulated.

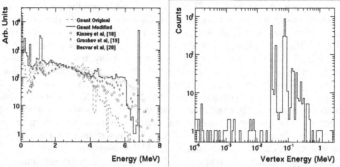

Figure 1: Left) Gamma energy spectra produced from neutron capture on ^{157}Gd for unmodified GEANT4 libraries, modified GEANT4 libraries and experimental data from [18-20]. The y-axis has been arbitrarily scaled to account for statistical differences. Right) The histogram presents the energy with which each electron is initially formed in modified GEANT4 simulations.

Table II: GEANT4 physics processes simulated for each specified particle; LE and HP correspond to Low Energy and High Precision, respectively.

Particle	Physics Processes
Neutron	HP Elastic & Inelastic Scattering, HP Capture, Neutron Induced Fission
Electron	Multiple Scattering, LE Ionization, LE Bremsstrahlung
Gamma	LE Rayleigh, LE Photoelectric, LE Compton, LE Conversion
Generic Ion	Multiple Scattering, LE Ionization

SIMULATION RESULTS

Analysis of 10^6 simulated neutrons reveals that electrons are produced in only 0.2% of events. This is expected because, although gadolinium has a high neutron capture cross section for the simulated energy range, Gd nuclei are present at a very low level in the diode. A histogram depicting the energy with which electrons are formed is shown in **Figure 1** (right). Peaks are observed in the distribution at 180, 79, 72, 39 and 30 keV. These correspond to the energies of internal conversion and Auger electrons produced during the de-excitation of ^{158m}Gd. The prevalence of these peaks confirms the expectation that the predominant reaction occurring in the material is neutron capture on ^{157}Gd despite its low abundance.

The fraction of energy deposited by each electron in the active volume of the diode can be determined from the simulation. It was found that only ~8% of electrons deposit their full energy in the detection region while on average electrons deposit ~20% of the energy with which they were created. Shown in **Figure 2** (left) is the total energy deposited by each electron in the sensitive region of the diode compared with the energy of electrons that experience full energy deposition. Electrons formed with energies >100 keV create little ionization as they exit the diode and contribute to the large low energy feature in the distribution. K shell Auger electrons with an initial energy of 30 keV have a low enough energy that they are fully captured in the active volume and are produced in sufficient quantities to result in a distinct peak in the electron energy spectrum. This peak remains distinguishable when the energy deposited per event by all particles is considered, **Figure 2** (right).

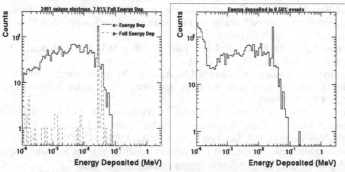

Figure 2: Left) The solid black histogram is the total energy deposited by each electron in the active volume of the diode. The dashed red histogram is the energy of the electrons that deposit their full energy in the active detector. Right) The histogram depicts the total energy per event deposited in the active volume of the diode by any particle species.

CONCLUSIONS

A GEANT4 modeling approach has been presented to evaluate the potential performance of solid-state neutron detectors incorporating intrinsic gadolinium. Strong motivation exists for developing new technologies capable of detecting neutron sources. Gadolinium possesses interesting nuclear and chemical properties that result in a high neutron capture cross section and the capability to be included in semiconducting devices. The utility of such devices to neutron detection applications is presently limited by the relatively thin layers incorporating gadolinium and the long range of the electrons with respect to the width of the depletion region. However, based on the simulation results, the diodes reported in [6,8,9] show the potential for detecting the 30 keV Auger electron from the decay of [158m]Gd provided that sufficient sensitivity can be achieved. This will require sensitive (1-10 fF) charge to voltage amplifiers, low noise readout and low device leakage currents. As reported in [8,9] these experimental conditions have not yet been achieved. Future computational studies should examine the effect of a high-energy gamma background present in high flux neutron testing environments.

ACKNOWLEDGMENTS

This work was supported by the Department of Homeland Security (IAA:HSHQDC-08-X-00641/P00001). The views expressed in this article are those of the authors and do not reflect the official policy or position of the Air Force, Department of Defense or the U.S. Government.

REFERENCES

1. D. Kramer, "DOE begins rationing helium-3", *Physics Today*, 22-25, June 2010.
2. M.B. Chadwick, P. Oblozinsky, M. Herman at al., "ENDF/B-VII.0: Next Generation Evaluated Nuclear Data Library for Nuclear Science and Technology", Nuclear Data Sheets, vol. 107, pp. 2931-3060, 2006.
3. Z.G. Ge, Y.X. Zhuang, T.J. Liu, J.S. Zhang, H.C. Wu, Z.X. Zhao, H.H. Xia, "The Updated Version of Chinese Evaluated Nuclear Data Library (CENDL-3.1)", Proc. International Conference on Nuclear Data for Science and Technology, Jeju Island, Korea, April 26-30, 2010 (in press).
4. I. Ketsman, Y.B. Losovyj, A. Sokolov, J. Tang, Z. Wang, M.L. Natta, J.I. Brand, P.A. Dowben, Appl. Surf. Sci. 254, 4308 (2008).
5. Y.B. Losovyj, D. Wooten, J.C. Santana, J.M. An, K.D., Belashchenko, N. Lozova, J. Petrosky, A. Sokolov, J. Tang, W. Wang, N. Arulsamy, P.A. Dowben, J. Phys. Condens. Matter 21, 045602 (2009).
6. D. Schultz, B. Blasy, J.C. Santana, C. Young, J.C. Petrosky, J.W. McClory, D. LaGraffe, J.I. Brand, J. Tang, W. Wang, N. Schemm, S. Balkir, M. Bauer, I. Ketsman, R.W. Fairchild, Y.B. Losovyj, P.A. Dowben, J. Phys. D: Appl. Phys. 43, 075502 (2010).
7. Monte-Carlo Simulation of Electron Trajectory in Solids (CASINO), accessed August 2009. Available: http://www.gel.usherbrooke.ca/casino/index.html.
8. B. Thomas, "Neutron Detection Using Gadolinium-Based Diodes," MS thesis, Department of Engineering Physics, Air Force Institute of Technology, Wright-Patterson AFB, OH, 2011.
9. C. Young, "Gadolinium Oxide / Silicon Thin Film Heterojunction Solid-State Neutron Detector," MS thesis, Department of Engineering Physics, Air Force Institute of Technology, Wright-Patterson AFB, OH, 2010.
10. S. Agostinelli, *et al.*, Nucl. Instrum. Methods Phys. Res. A 506, 250 (2003).
11. R.G. Helmer, Nuclear Data Sheets 101, 325 (2004).
12. GEANT4 Collaboration, "Data files for photon evaporation - version 2.1", released February 2011. Available: http://geant4.web.cern.ch/geant4/support/download.shtml
13. M.A. Ali, V.A. Khitrov, Y.V. Kholnov, A.M. Sukhovoj, A.V. Vojnov, J. Phys. G Nucl. Part. Phys. 20, 1943 (1994).
14. M.A. Islam, T.J. Kennett, W.V. Prestwich, Phys. Rev. C 25, 3184 (1982).
15. R.B. Firestone, H.D. Choi, R.M. Lindstrom, G.L. Molnar, S.F. Mughabghab, R. Paviotti-Corcuera, *et al.*, Database of prompt gamma rays from slow neutron capture for elemental analysis, (Lawrence Berkeley National Laboratory, California, 2004) p. 132.
16. GEANT4 Collaboration, "Physics Reference Manual geant4.9.4", released December 2010, p 499. Available: http://geant4.web.cern.ch/geant4/UserDocumentation/UsersGuides/PhysicsReferenceManual/fo/PhysicsReferenceManual.pdf
17. F. Rosel, H.M. Fries, K. Alder and H.C. Pauli, At. Data Nucl. Data Tables 21 (1978).
18. Kinsey, *et al.*, Can. J. Phys. 31, 1051 (1953).
19. Groshev, *et al.*, Atomnaya Energiya 4, No 1, 5 (1958).
20. F. Becvar, M. Krticka, I. Tomandl, J. Honzatko, F. Voss, K. Wisshak, F. Kappeler, AIP Conf. Proc. 529, 657 (2000).

Mater. Res. Soc. Symp. Proc. Vol. 1341 © 2011 Materials Research Society
DOI: 10.1557/opl.2011.1481

Material parameter basis for major and minor trends in nonproportionality of scintillators

Qi Li, Joel Q. Grim, R. T. Williams
Department of Physics, Wake Forest University, Winston-Salem, NC 27109
G. A. Bizarri, W. W. Moses
Lawrence Berkeley National Laboratory, Berkeley, CA 94720

Abstract
    We have previously described a numerical model for carrier diffusion and nonlinear quenching in the track of an electron in a scintillator. Significant inequality of electron and hole mobilities predicts a characteristic "hump" in the light yield vs gamma energy, whereas low mobility of either or both carriers accentuates the universal roll-off due to nonlinear quenching at low gamma energy (high dE/dx). The material parameter basis of the two major trends in nonproportionality of scintillators can be related to the effective diffusion coefficient of excitations and the difference of electron and hole mobilities, respectively. Activator concentration, type of activator, and effect of transport anisotropy are associated with minor trends. The predicted trends are qualitatively consistent with empirical measures of nonproportionality including electron yield curves.

Keywords: Scintillators, nonproportionality, carrier diffusion, mobility, nonlinear quenching

INTRODUCTION

    Intrinsic nonproportionality is a material dependent phenomenon that sets an ultimate limit on energy resolution of radiation detectors. In general, anything that causes light yield to change along the particle track (e.g., the primary electron track in γ-ray detectors) contributes to nonproportionality. It is widely accepted that nonlinear quenching is a root cause of scintillator nonproportionality. However, the question remains: "What are the independently measurable and calculable material parameters that control nonlinear quenching?"

    Payne et al [1] have fit electron yield data from the SLYNCI (Scintillator Light Yield Nonproportionality Characterization Instrument) experiment for a number of scintillator materials using two empirical fitting parameters: a "Birks parameter" meant to characterize how strong the 2nd order dipole-dipole quenching term is, and a fraction $\eta_{e/h}$ of the initial electron-hole excitations that branch into independent carriers rather than excitons. They have investigated a number of material trends in SLYNCI electron yield data including dependence on the host crystal characteristic, activator concentration, and activator type. [1,2]

    We have recently presented a numerical model showing how mobilities μ_e, μ_h [3,4], and corresponding diffusion coefficients for electrons, holes and excitons D_e, D_h, D_{EXC} [5] give a quantitative account of nonproportionality in oxide scintillators and semiconductor radiation detectors, and a qualitative account of characteristic trends of light yield in halide scintillators. In this paper, we examine whether the empirical trends in nonproportionality and energy resolution described in Refs. [1 and 6] can be described by the transport and nonlinear quenching model.

    Setyawan et al have investigated the link between band effective masses and nonproportionality [6]. Faced with the scarcity of measured mobilities or effective masses for most scintillators, they took the course of calculating electronic band structure for a wide range of scintillators in order to deduce effective masses from the band curvatures. In this paper, we will in some cases scale the mobility of carriers from their calculated band masses.

TRANSPORT MODEL

The method of numerical modeling of nonlinear quenching and transport in electron tracks has been described in Refs. [3-5]. We plot in Fig. 1 the simulations of 1-QF, which is the normalized fraction of electron-hole pairs surviving 2nd order quenching in CsI at the typical quenching time window 10 ps. For this plot, electron and hole diffusion coefficients are assumed to be equal.

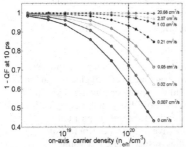

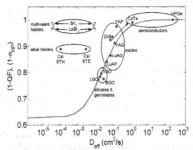

Fig. 1 Probability of survival against 2nd order quenching evaluated at 10 ps, plotted as a function of on-axis excitation density in an electron track deposited in materials having the electron and hole diffusion coefficients listed at the right of each curve.

Fig. 2 The curve (see text) represents the ability of excitations to escape from the track by diffusion before nonlinear quenching can occur. Empirical measures of proportionality (1- σ_{NP}) [6] are plotted versus effective diffusion coefficient, D_{eff} from Eq. (1).

The vertical dashed line in Fig. 1 marks the excitation density of 10^{20} e-h/cm^3, approximately half of the electron-hole density at the track end. The intersection of this line with the curves 1-QF for different diffusion coefficients spanning 8 decades are plotted as the curve in Fig. 2. In Ref [5], we regarded the excitations in a track as a mixture of free carriers in the relative fraction IF, and of excitons in the relative fraction 1-IF. The exciton fraction 1-IF was modeled as linearly proportional to the similarity of the electron and hole effective masses, as

$$1 - IF \approx \min(\frac{m_e}{m_h}).$$ The exciton should experience less scattering by charged defect and optical phonons than the independent carriers, and thus its diffusion coefficient should be proportional to a longer momentum relaxation time τ_{EXC} than that of a free carrier, taken as $\tau_e \approx \tau_h$. The effective diffusion coefficient was expressed as a linear combination of exciton and free carrier contributions in the following way,

$$D_{eff} \approx \frac{kT}{\max(m_h, m_e)} \left[\min\left(\frac{m_e}{m_h}\right)\tau_{EXC} + \left(1 - \min\left(\frac{m_e}{m_h}\right)\right)\tau_e \right] \qquad (1)$$

With the ability to calculate D_{eff} for each material, we can superimpose experimental measurements of nonproportionality on the theoretical curve in Fig. 2. The experimental nonproportionality parameter σ_{NP} was compiled by Setyawan et al [6]. In Fig. 2, we plot the proportionality parameters 1-σ_{NP} for all of the oxides and for ZnSe from the tabulation of Setyawan et al [6], along with the proportionality parameters for CdTe and HPGe. The agreement for the oxides and semiconductors is remarkable. There were no adjustable parameters for the curve or the data except for the choice of τ_{EXC}.

Taking CsI:Tl for example, independent trapped charges undergo de-trapping and re-trapping process until they recombine as Tl^{+*} in order to yield luminescence. This is a different recombination path with different time dependence (corresponding with the slow component) and different perils for quenching or deep-trapping than in the alternate path take by electron and hole initially trapped as a pair on the same Tl^+ (corresponding with the fast component). We incorporate the trapping hazard for migrating free carriers in the model as a linear quenching fraction k_l that multiplies the independent carrier fraction IF to give the "Independent Nonradiative Fraction", INF:

$$INF = k_l \times IF \qquad (2)$$

Let us assume that the probability of the independent carriers to survive for a single deep-trapping event is σ. The probability to survive for the whole recombination path of length l defines the linear quenching fraction $k_l(l)$ dependent on activator concentration through l as:

$$P(l) = 1 - k_1(l) \propto (1-\sigma)^{\frac{l}{s_0}} \qquad (3)$$

where l is proportional to the activator spacing, and s_0 is the spacing of defects responsible for linear quenching.

In the context of this model, we define "simulated local light yield" ($SLLY$) as follows:

$$SLLY(normalized) = (1 - QF)(1 - INF) \qquad (4)$$

It predicts an upper limit of the local light yield as a function of initial carrier concentration. In Fig. 3, we plot $SLLY$ to represent the approximate combined probability of an excitation surviving both nonlinear dipole-dipole quenching and conversion to independent carriers with exposure to linear quenching. Although not an actual electron yield curve, the local light yield versus excitation density should have qualitative similarity to electron yield curves. Notice from Fig. 3 that the hump becomes most pronounced when D_h/D_e is very small, as is realized only with deep hole self-trapping found in the alkali halides. Payne et al [1] previously described how the hump can empirically flatten the electron yield curve over part of its range. A particularly large hump as in the alkali halides will impose a proportionality cost of slope on the way up and on the way down. Therefore as a qualitative goal, a modest hump that puts the flat slope in a good place without introducing big slopes up and down would be ideal in this regard. Our model predicts that some of the multivalent halides with anisotropic electronic structure have the light yield response like this, which suggests a reason for their good proportionality.

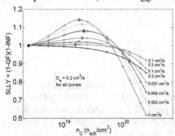

Fig. 3 Simulated local light yield $(1-QF)(1-INF)$ is plotted for each hole diffusion coefficient listed on the right of each curve, while keeping D_e at the CsI value. Increasing D_h corresponds monotonically to increasing height of the "hump" in this simulated local light yield versus on-axis excitation density (proportional to dE/dx).

EMPIRICAL TRENDS COMPARED TO TRANSPORT MODEL PREDICTIONS

A. The host crystal characteristic

Payne et al found in a survey of electron yield in scintillators that the strongest trend is a grouping according to host lattice.[1,2] This immediately raises the question of what is the material parameter that changes according to electronic structure or host lattice and exerts a strong effect on nonproportionality. Based on the transport model reviewed in the previous section, we submit that the effective diffusion coefficient of excitations is the responsible parameter. This should be understood to include both of the effects discussed above, namely, the ability to escape the quenching zone in the track and the control over exciton and free carrier branching exerted by relative diffusion coefficient. To illustrate that Fig. 2 amounts to the trend noted by Payne et al, we have circled the groups of materials with similar structures. It can be seen that the semiconductors, oxides and silicates are in descending order of both proportionality and effective diffusion coefficient. The halides break that trend between effective diffusion coefficient and proportionality, but the probable reason was illustrated already in Fig. 3, based on inequality of diffusion coefficients for electrons and holes.

B. Activator concentration

While measurements of light yield dependent on primary electron energy from the SLYNCI experiments [1,2] showed substantial dependence on host crystal structure (see above), only weak effects of the concentration of activator ions were found. SLYNCI electron yield curves for LaBr$_3$:Ce measured in the very large range of Ce concentrations 0.5%, 5%, 10%, 20%, 30% exhibited only slight variation.[2] The model reviewed in the previous section included treatment of a linear quenching fraction k_1 in which activator concentration is a factor. But as shown in Refs. [3-5], the independent carrier fraction IF in that model also depends on activator concentration, and the two dependences approximately cancel in the product k_1 IF. We have performed simulations for the same series of concentrations of LaBr$_3$:Ce as had been run in the SLYNCI experiments. The simulated local light yields are shown in Fig. 4. The shape and its (near lack of) dependence on Ce concentration is similar to what was found in the experiment [4].

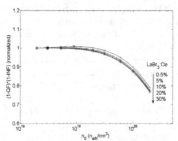

Fig. 4 Simulated local light yield of LaBr$_3$:Ce with different activator concentration as a function of on-axis carrier density. The weak dependence on cerium activator concentration is similar to experimental observations in SLYNCI electron yield.[2]

C. Activator type: CsI:Tl versus CsI:Na

SLYNCI electron yield data show dependence of light yield on activator type in the comparison of CsI:Tl and CsI:Na. CsI:Na has a bigger "hump" than CsI:Tl [2]. Parameters of our transport model that can be considered characteristic of a particular activator are the time for trapping charges on the activators and the multi-valency, or the ability to trap either one or both charges. Fig. 5 shows the dependence on the trapping time. We can see that an increasing time for trapping increases the height of the hump modestly. An alternative qualitative explanation is related to the recombination process. In CsI:Tl, there are 3 channels of luminescence, the exciton trapping followed by direct radiation, the prompt electron capture forming Tl^0 followed by diffusion of V_k centers to Tl^0 to form Tl^{+*}, and the electron hopping from Tl^0 followed by their diffusion to Tl^{++} formed by the V_k center capture. Both the 2nd and 3rd channel are related to the independent non-radiative fraction, since production of independent carriers is favored at the beginning of the track when mobilities are mis-matched. This means light yield is limited early in the track, suggesting a reason for the rising part of light yield in alkali halides. In contrast, Na atoms can only lose one electron, which means the 3rd channel is prohibited in CsI:Na. This fact leads to a less effective recombination for independent carriers in CsI:Na. To state in another way, the average distance for the independent trapped carriers to recombine is increased. Hence a greater value of linear quenching fraction k_l due to intervening defects is expected in CsI:Na. In Fig. 6, we plot the simulated local light yield with changing k_l, showing that a higher "hump" occurs for greater k_l, which is consistent with the comparison in SLYNCI experiments. [1,2].

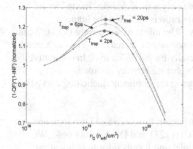
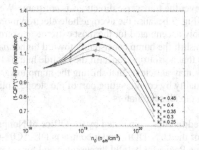

Fig. 5 *SLLY* plotted versus on-axis density for CsI with varying trapping times listed in the legend.

Fig. 6 simulated local light yield is plotted versus on-axis carrier density, each curve is for a different value of linear quenching fraction k_l.

D. Alkali (cubic) halides vs. multivalent (non-cubic) halides

It was pointed out in Ref. [4] that the group of multivalent halides has higher proportionality than the alkali halides, or monovalent halides. We know that all of the alkali halides are cubic, while many of the multivalent halides are non-cubic; for example, SrI_2 has orthorhombic structure and $LaBr_3$ has hexagonal structure. Therefore, it is worthwhile to explore the effects of anisotropy in the transport properties of scintillators and its predictions for nonproportionality.

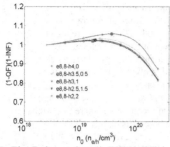

Fig. 7 Simulated local light yield for five assumed degrees of anisotropy in hole mobility, keeping the average hole mobility constant along with the isotropic electron mobility. The notation e8,8-h4,0 means, e.g., that the electron mobility is 8 cm^2/Vs in two orthogonal directions transverse to the track, and that the hole mobility is 4 cm^2/Vs on the X transverse axis and 0 cm^2/Vs on the Y axis.

In Fig. 7, the product $(1\text{-}INF)(1\text{-}QF)$ is plotted for the indicated combinations of variable anisotropic hole mobility and fixed isotropic electron mobility, where the average of hole mobility over the transverse directions is constant. We see only a weak halide "hump" in the curves of Fig. 7 because the average hole/electron mobility ratio is 0.25, but the hump is considerably accentuated for the most extreme anisotropy. The effect of anisotropy in Fig. 7 can be seen to shift the hump consistently toward higher n_0 as anisotropy increases. Thus as a trend, it appears that anisotropy can shift the "halide hump" laterally on the dE/dx axis. In qualitative terms, one may anticipate that shifting the hump toward higher dE/dx may improve proportionality by counteracting part of the steepest plunge in the nonlinear quenching curve that is common to all scintillators.

Acknowledgments-- supported by the Office of Nonproliferaton Research and Development (NA-22) of the U. S. DOE, under contracts DE-AC02-05CH11231 and DE-NA0000668. We thank Steve Payne for helpful discussions and for communicating results prior to publication.

REFERENCES
1. S. A. Payne, N. J. Cherepy, G. Hull, J. D. Valentine, W.W. Moses, and Woon.-Seng. Choong, IEEE Trans. Nucl. Sci. **56**, 2506 (2009).
2. S. A. Payne, W. W. Moses, S. Sheets, L. Ahle, N. J. Cherepy, B. Sturm, S. Dazeley, private communication.
3. Qi Li, Joel Q. Grim, R. T. Williams, G. A. Bizarri, and W. W. Moses, Nucl. Instrum. Methods Phys. Res. A (2011), doi:10.1016/j.nima.2010.07.074.
4. R. T. Williams, J. Q. Grim, Qi Li, K. B. Ucer, and W. W. Moses, "Excitation density, diffusion-drift, and proportionality in scintillators", Phys. Status Solidi B, **248**, 426 (2011).
5. Qi Li, Joel Q. Grim, R. T. Williams, G. A. Bizarri, and W. W. Moses, J. Appl. Phys., **109**, 123716 (2011).
6. W. Setyawan, R. M. Gaume, R. S. Feigelson, and S. Curtarolo, IEEE Trans. Nucl. Sci., **56**, 2989 (2009).

Mater. Res. Soc. Symp. Proc. Vol. 1341 © 2011 Materials Research Society
DOI: 10.1557/opl.2011.1273

Dimensionally reduced heavy atom semiconductors as candidate materials for γ-ray detection: the case of $Cs_2Hg_6S_7$

Ioannis Androulakis,[1] Hao Li,[1] Christos Malliakas,[1] John A. Peters,[2] Zhifu Liu,[2] Bruce W. Wessels,[2,3] Jung-Hwan Song,[4] Hosub Jin,[4] Arthur J. Freeman,[4] Mercouri G. Kanatzidis[1]

[1]Department of Chemistry, Northwestern University, Evanston, IL, 60208, USA
[2]Department of Materials Science and Engineering, Northwestern University, Evanston, IL, 60208, USA
[3]Department of Electrical Engineering, Northwestern University, Evanston, IL, 60208, USA
[4]Department of Physics and Astronomy, Northwestern University, Evanston, IL 60208, USA

ABSTRACT

We address the issue of decreasing band-gap with increasing atomic number, inherent in semiconducting materials, by introducing a concept we call dimensional reduction. The concept leads to semiconductor compounds containing high atomic number elements and simultaneously exhibiting a large band gap and high mass density suggesting that dimensional reduction can be successfully employed in developing new γ-ray detecting materials. As an example we discuss the compound $Cs_2Hg_6S_7$ that exhibits a band-gap of 1.65eV and mobility-lifetime products comparable to those of optimized $Cd_{0.9}Zn_{0.1}Te$.

INTRODUCTION

Currently, interest in semiconductors for hard radiation detection for home-land security and nuclear non-proliferation applications is increasing [1]. The elemental and compound semiconductors used in applications or under development have several limiting disadvantages. For example, high purity Ge requires cryogenic cooling under operating conditions because of its low band gap (~0.6 eV) [1, 2], and $Cd_{1-x}Zn_xTe$, the leading compound for room-temperature operation, faces challenging issues such as Te particle precipitation, scale up and an energy gap, E_g~1.57 eV, which is somewhat lower than desired giving rise to noticeable dark current [3, 4].

Achieving coexistence of a high atomic number, Z, high mass density and large E_g values in semiconductor crystals is a challenging problem. Increasing Z leads to increased overlap of wavefunctions strongly contributing to band broadening and low band gaps [5]. The E_g-Z relation imposes severe limitations on the number of existing candidate compounds for hard radiation detection.

In order to find better materials with enhanced radiation detection capabilities, the limiting E_g-Z relation has to be effectively addressed. In this work we present a concept we call dimensional reduction to show that Hg containing compounds can form the basis for generating large E_g materials without compromises on the mass density. We report the synthesis, crystal growth and characterizations as well as band-structure calculations of $Cs_2Hg_6S_7$, which comprises of HgS dimensionally reduced with Cs_2S. The results are comparable to current state-of-the-art compounds such as $Cd_{1-x}Zn_xTe$ suggesting that dimensional reduction is a promising method to realize future γ-ray detection materials.

EXPERIMENTAL AND THEORETICAL METHODS

High purity Hg (99.9999%, Aldrich), Cs (>99.9, Aldrich), and S (99.98%, 5N Plus Inc.) were used as starting materials. Cs_2S was synthesized in liquid ammonia and HgS was synthesized by transport of Hg and S vapors inside a sealed quartz crucible. Bulk polycrystalline ingots of the end material were produced at 700 C according to the reaction scheme: Cs_2S + $6HgS \rightarrow Cs_2Hg_6S_7$. The polycrystalline powders were used for crystal growth in home-made furnaces employing the Bridgman-technique.

Density functional theory calculations were performed using the full-potential linearized augmented plane wave (FLAPW) [6] method with the screened-exchange local density approximation (sX-LDA) [7], which is known to yield great improvements of the excited electronic states over the conventional LDA. Thus estimates of band gaps and band dispersions are better approximated. The core and valence states were treated fully relativistically and scalar relativistically, respectively. The spin-orbit coupling is also included by employing a second variational procedure [8]. The lattice parameters values and atomic coordinates of single crystal diffraction refinements were used in the calculations [9]. For k-space integrations, a 3x3x9 mesh of special k-points was used in the irreducible Brillouin zone wedge. The energy cutoffs for the interstitial plane-wave basis and the star functions were 12.3 and 144 Ry, respectively.

Powder x-ray diffraction patterns were collected with an INEL diffractometer equipped with a Cu Ka radiation. The electrical resistivity was measured at room temperature with a 4-probe method using a Keithley 2182A nano-voltmeter and a Keithley 6220 current source. A Shimadzu UV-3101PC spectrophotometer was used to define the band gap at 300 K.

The mobility-lifetime products for electrons and holes were determined by photoconductivity measurements and compared to the values of $Cd_{0.96}Zn_{0.04}Te$ research grade single crystal purchased from MTI Corporation. The photoconductivity measurements setup comprised of a He-Ne laser light source at a wavelength of 633 nm, filtered and focused with a lens, a chopper operating at 200 Hz, a high voltage power supply, and a lock-in amplifier.

DISCUSSION

The concept of dimensional reduction

Previously we have shown that lowering the dimensionality of a compound semiconductor structure creates a blue shift in the energy gap [9, 10]. Here, we call dimensional reduction the process by which this lowering takes place in a controllable fashion by incorporation of Cs_2S into the binary network of HgS. The former amounts to the introduction of S^{2-} atoms in the cubic lattice of the binary host and the generation of anionic $[Hg_xS_y]^{n-}$ frameworks with a lower dimensionality [9, 10]. The lattice of HgS is forced to accommodate the extra S^{2-} atoms which progressively dismantle it (i.e. reduce its dimensionality). The Cs^+ cations act as charge balancing spacers that interact only with electrostatic forces with the $[Hg_xS_y]^{n-}$ frameworks and therefore do not significantly perturb the band structure of HgS, that depends only on the changing framework dimensionality. This way we can create chemical homologies of the type $(Cs_2S)_m(HgS)_n$. Theoretical band-calculations can then be used to aid in the selection processes of promising compounds within a certain homology.

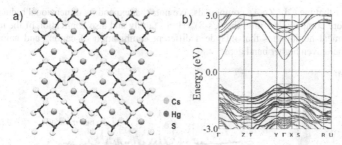

Figure 1: a) The crystal structure of $Cs_2Hg_6S_7$ (space group $P\,4_2nm$). b) Corresponding DFT electronic band structure calculation.

The case of $Cs_2Hg_6S_7$

In the family of compounds $(Cs_2S)_m(HgS)_n$ theoretical calculations indicated that the material with m=1 and n=6 is a potentially promising compound for radiation detection. A discussion on other members of the above family of compounds is not possible here due to space limitations however the issue will be addressed in a future publication. $Cs_2Hg_6S_7$ exhibits a three-dimensional tetragonal network of Hg and S, Fig 1a. The band structure of both conduction and valence bands becomes more dispersive along the Γ to Z direction as depicted in Fig. 1b. Both the conduction and valence bands show dispersive bands near the band edges suggesting light n- and p-type carriers.

RESULTS

Figure 2a shows a typical single-crystalline boule grown with the Bridgman method. During growth the hot zone was operated ~50 °C above the melting point of the compound (586 °C) and the cold zone was operated ~40 °C below the recrystallization point (542 °C). The melting and recrystallization points were defined with differential thermal analysis. Parts of the grown crystal were pulverized and subsequently used in further characterizations. Figure 2b depicts an x-ray powder diffraction pattern received from such pulverized specimens. The experimental diffractogram is compared to a reference pattern that was generated based on the space group, lattice parameters and electron densities of the constituent elements of $Cs_2Hg_6S_7$ (s.g. $P\,4_2nm$, a=b=14.505 Å, c=4.308 Å). The agreement is excellent, suggesting high purity within the resolution of the instrument. Extensive back-reflection Laue patterns were also received from different parts of the boule to ensure its single-crystalline nature.

Figure 2c presents diffuse reflectance measurements as a function of energy on pulverized chunks of the single-crystalline boule. BaSO4, was used as 100% reflectance standard material, preloaded into a sample holder. The bandgap was determined as the intersection point between the energy axis and the line extrapolated from the linear portion of the absorption edge of the Kubelka-Munk analyzed diffuse reflectance spectra, R, i.e. $(1-R)^2/2R$ vs $\hbar\omega$, were ω is the frequency and $\hbar$ Planck's constant [11]. The band gap was determined to be ~1.65 eV at room-temperature in rough agreement with the theoretical band structure calculations, Fig. 1b. This value is higher that the $Cd_{1-x}Zn_xTe$ compounds. We note that expanding dimensional reduction to synthesize more compounds within the $(Cs_2S)_m(HgS)_n$ family and also application of the method to other chemical homologies that include a different combination of cations and anions are expected to yield even higher band gaps.

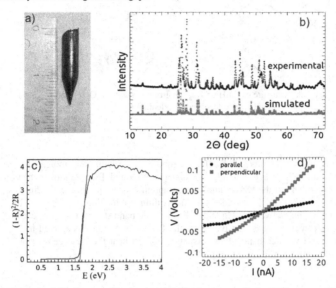

Figure 2: a) Single-crystalline boule of $Cs_2Hg_6S_7$ grown with the Bridgman method. b) Powder x-ray diffraction pattern from pulverized parts of the boule shown in a) compared to the simulated pattern based on the space group and lattice parameters. c) Diffuse reflectance as a function of energy at 300K. The band gap is ~1.65 eV. d) 4-probe I-V curves at 300 K parallel (black circles) and perpendicular (red squares) to the growth directions. The extracted resistivities are ~ $2x10^6$ Ω·cm and $7x10^6$ Ω·cm respectively.

A high electrical resistivity, pertaining to high purity, is important in detector performance [1, 12]. A large dark current result in peak broadening by leakage noise. Current-Voltage (I-V) characteristic curves for two specimens cut parallel and perpendicular (ab plane) to the growth direction are presented in Fig. 2d. The electrical resistivity calculated for $Cs_2Hg_6S_7$ is ~$2x10^6$ Ω·cm for the specimen in the growth direction and ~$7x10^6$ Ω·cm in the direction perpendicular to the growth. This is consistent with the 3-D nature of the crystal structure of the compound. The resistivity values are lower when compared to current state-of-the-art $Cd_{1-x}Zn_xTe$, but improvements are expected with further development.

In order to determine $\mu\tau$ for both electrons and holes photo-conductivity measurements were performed and data were analyzed according to the method proposed by Many et al. [13]. Different voltage polarities were applied to the illuminated electrode so that the photocurrent for electrons and holes was separately measured, and their corresponding $\mu\tau$ products obtained. For strongly absorbed light the observed photo-current can be expressed as a function of the applied biased voltage, V [13]:

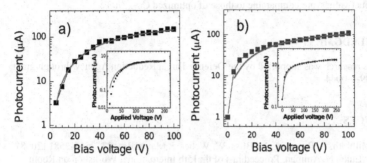

Figure 3: Photo-conductivity measurements in $Cs_2Hg_6S_7$ for a) electrons and b) holes. The red solid line is an iterative fit of the data using eq. 1 (see text) which is used to determine the $\mu\tau$ product.

$$I(V) = \frac{I_0 \mu\tau V}{L^2} \frac{\left(1 - e^{\frac{-L^2}{\mu\tau V}}\right)}{1 + \frac{L}{V}\frac{s}{\mu}}$$

(1)

where I_0 is the saturation value of the photo-current, L is the sample thickness, and s the surface recombination velocity. A fit of the observed photo-current to eq 1 yields both $\mu\tau$ and s/μ. Using eq. 1 and reversing bias voltage $\mu\tau$ and s/μ can be extracted both for electrons and holes. The value of the ratio s/μ can be used as an evaluation criterion for surface quality. A low quality crystal surface is expected to increase unwanted recombination effects and thus limit the value of $\mu\tau$ of the bulk. Figure 3 a-b presents photocurrent as a function of bias voltage for a $Cs_2Hg_6S_7$ z-cut single crystal, where z is the direction of the growth. The red solid line in all cases is the best iterative fit of the data to eq. 1. The $\mu\tau$ values for $Cs_2Hg_6S_7$ electrons were determined $\sim 2 \times 10^{-3}$ cm^2V^{-1} while for holes $\sim 3.4 \times 10^{-4}$ cm^2V^{-1}. These values are consistent with the DFT calculations that predict a higher dispersion of the conduction band and hence a lower electronic mobility due to decreased effective mass values compared to holes.

To check the validity of our measurements for the $\mu\tau$ determination of $Cs_2Hg_6S_7$ we applied our measurement protocol to a commercial research grade $Cd0.96Zn_{0.04}Te$ single-crystal purchased from MTI corporation. The derived $\mu\tau$ values were in agreement with literature values measured on optimized $Cd_{0.9}Zn_{0.1}Te$ [12]. The double comparison of $Cs_2Hg_6S_7$ $\mu\tau$ values to state of the art literature values on one hand [12] and to an actual commercial $Cd_{0.96}Zn_{0.04}Te$ crystal on the other underscores the promising properties of our compound.

In conclusion, dimensional reduction constitutes a promising method in designing novel materials for γ-ray detection. We have presented one example of dimensionally reduced HgS through Cs_2S to form $Cs_2Hg_6S_7$. The Hg rich tetragonal lattice of $Cs_2Hg_6S_7$ exhibits a band gap of 1.65eV and a high mass density (6.94 g/cm^3). Despite a low electrical resistivity, ~10^6 Ωcm, the $\mu\tau$ product values are comparable to those of optimized $Cd_{1-x}Zn_xTe$.

ACKNOWLEDGMENTS

This research was supported by the Defense Threat Reduction Agency through grant HDTRA1 09-1-0044.

REFERENCES

1. B. Milbrath, A. Peurrung, M. Bliss, W. Weber, J. Mater. Res. **23**, 2561-2581 (2008).
2. P.N. Luke, N. Amman, Proceedings of the15th International Workshop on Room-Temperature Semiconductor X- and Gamma-Ray Detectors, San Diego, CA, 2006, pp. 834.
3. T.E. Schlesinger, J.E. Toney, H. Yoon, E.Y. Lee, B.A. Brunett, L. Franks, R.B. James, Mater. Sci. Eng., R **32**, 103 (2001).
4. A. Bolotnikov, G. Camarda, G. Wright, R. James, IEEE Trans. Nucl. Sci. **52**, 589 (2005).
5. B.R. Nag, Infrared Phys. Technol. **36**, 831 (1995).
6. Wimmer, E., Krakauer, H., Weinert, M., and Freeman, A. J., *Physical Review B* **24** (2), 864 (1981).
7. Bylander, D. M. and Kleinman, L., *Physical Review B* **41** (11), 7868 (1990).
8. A. H. MacDonald, W. E. Pickett, and D. D. Koelling, *J. Phys. C* **13**, 2675 (1980).
9. E.A. Axtell, Y. Park, K. Chondroudis, M.G. Kanatzidis, J. Am. Chem. Soc. **120**, 124 (1998)
10. E.A. Axtell, J. Liao, Z. Pikramenou, M.G. Kanatzidis, Chem. Eur. J. **2**, 656 (1996).
11. P. Kubelka, Z. Munk, Z. Tech. Phys. **12**, 593 (1931).
12. A. Owens, A. Peacock, Nucl. Instrum. Methods Phys. Res. Sect. A **531**, 18-37 (2004).
13. A. Many, High-field effects in photoconducting cadmium sulphide, *J. Phys. Chem. Solids* **26**, 575 (1965).

Other Materials

Mater. Res. Soc. Symp. Proc. Vol. 1341 © 2011 Materials Research Society
DOI: 10.1557/opl.2011.1111

Czochralski Growth of Indium Iodide and other Wide Bandgap Semiconductor Compounds

I. Nicoara[1], D.Nicoara[1], C. Bertorello[1], G.A. Slack[1] and A. G. Ostrogorsky[1], M. Groza[2] and A. Burger[2]

[1]Illinois Institute of Technology (IIT), Chicago, IL 60616
[2]Fisk University, Physics Department, Nashville, TN 37208

ABSTRACT

The Czochralski pulling process is the most valuable and cost efficient method for producing large oriented single crystals of the group IV and III-V semiconductors. However, there have been only a small number of reported attempts to use the Czochralski process for growing the wide bandgap compound semiconductors, needed for the room temperature operated gamma-ray detectors. The main difficulty is in the low chemical stability and high vapor pressure of the group II, V and VI elements, leading to off-stoichiometric composition, and various related defects. Among the heavy metal halides, indium iodide and indium bromide present an interesting exception. InI has a high molecular disassociation energy and a low vapor pressure, allowing for Czochralski pulling. We will describe the procedures used and the results obtained by Czochralski growth and characterization of indium iodide and the related ternary compounds that appear to be quite encouraging.

INTRODUCTION

The Czochralski crystal pulling, from the free surface of the melt, is the most valuable and cost efficient method for producing large oriented single crystals. The well established advantages of the Czochralski growth process include:

   i. The crystals grow and cool down unrestricted by the crucible walls.
   ii. Forced convection is easy to impose.
   iii. High throughput; large crystals can be obtained.
   iv. The growing crystal can be observed.
   The main drawbacks, compared to growth in sealed ampoules, are:

   a. The high vapor pressure materials have to be grown under liquid encapsulation using the Liquid Encapsulated Czochralski (LEC) process, or high inert gas pressure, in high pressure CZ pullers.
   b. The process requires continuous attention during seeding and necking
   c. For high melting point materials, temperature gradients are relatively high.

So far, only B_2O_3 has proven to be a useful encapsulant used for LEC growth of the III-V arsenides and phosphides. There has been only a small number of attempts to use LEC for growing the wide bandgap II-VI compounds [1,2]. These later attempts have not been successful, because apparently, B_2O_3 reacts with the melt. Furthermore, the group II and VI elements are both volatile, making stoichiometry control difficult and leading to various related defects.

A number of heavy metal iodides has been investigated as promising room temperature detector materials, because of their high density, stopping power, and useful energy band gap, Eg> 1.6 eV. These iodides include HgI_2, PbI_2, BiI_3, InI, TlBr and TlI. Note that the heavy metal iodides are all are layered materials, with the exception of TlBr which has the cubic BCC structure. Yet, there is a significant difference between:

- (a) the di- and tri-valence halides, built from three-layer sheets where the heavy metal atom is sandwiched between the halides (e.g. I-Hg-I) and the sheets are weakly bonded by the Van der Waals forces, and
- (b) the monohalides InI and TlI which have layers comprised of "dumbbell" structured molecules. The layers are held together by ionic-covalent bonds between In (or Tl) and iodine atoms, and therefore are much stronger than the van der Waals forces between iodine layers.

The diatomic molecules of indium mono-iodide are very stable compared to the molecules of the other heavy metal halides [3], see Table I. While InBr has the highest dissociation energy, InI was chosen for CZ growth because of its higher density, melting point and Z number. The interatomic separation of InI in the gas phase is 2.75 Å. These molecules also exist in the solid InI, which has a single orthorhombic crystal structure, where the interatomic distance is: 3.23 Å parallel to the c_0-axis ($a_0 = 4.73$ Å, $b_0 = 12.76$ Å, $c_0 = 4.91$ Å) see Wyckoff[4]. InI has solid mechanical properties and it is not toxic.

Single crystals of InI were grown in sealed ampoules for detector applications, at Radiation Monitoring Devices (RMD) [5] and more recently in our laboratory [6]. It was concluded that the commercial purity of the elements for the synthesis of InI is inadequate for semiconductor detector applications. The electrical resistance was drastically improved to 5×10^{11} Ωcm by 50 passes of zone refining [7].

Table I Data from [3]

	Disassoci-ation Energy , eV	Internuclear distance, 10^{-8}cm
I_2	1.542	2.7
BiI	0.3	
HgI	0.35	2.68
HgBr	0.71	2.57
CdTe	1.2	
PbI	2.0	3.10
PbBr	2.5	2.81
TlI	2.76	2.81
TlBr	2.34	2.62
InI	3.43	2.75
InBr	3.9	2.54

To our knowledge, there has been no attempts to use the Czochralski process to grow the above listed heavy metal halides, with the exception of PbI$_2$ [8].

EXPERIMENTAL DETAILS

Synthesis of InI

The compound melts congruently at 365 ° C, see the Indium-Iodine phase diagram as reviewed by Peretti [9]. The melt is in equilibrium with the gaseous phase of InI diatomic molecules. The vapor pressure at the melting point is 1.70×10^{-3} bars and approaches 1 bar at 732 ° C (see Barin's Tables, [3]). The residual pressure of I$_2$ gas over InI is much less than that of InI molecules. The gas pressure ratio is:

$$\frac{p(I_2)}{p(InI)} = 10^{-14.5} = 3.16 \times 10^{-15} \text{ at the melting point.}$$

$$\frac{p(I_2)}{p(InI)} = 10^{-9.85} = 1.41 \times 10^{-10} \text{ at the boiling point.}$$

Thus there is very little dissociation of InI solid or liquid into liquid In and I$_2$ gas. The synthesis of InI is best carried out in a two-zone furnace (figure 1), where In is held approximately 20 ° C above the melting point of InI, while iodine is at about 70 ° C where its vapor pressure is 10^{-2} bars. Low I$_2$ pressure prevents a runaway exothermic reaction. Liquid iodine, as I$_2$ liquid has a boiling point of 184.4 ° C. The starting charge should be approximately 1 % rich in indium, to prevent formation of higher iodides, InI$_2$ and InI$_3$, which have lower melting points and higher vapor pressures. The synthesis requires approximately 4 hours.

Peretti [9] has shown that liquid indium monoiodide and liquid indium are immiscible. Molten InI is lighter (ρ=5.4 g/cm^3) than molten indium (ρ=7.02 g/cm^3) for indium. Thus, any InI made from indium liquid and I$_2$ vapor will float on the top of the remaining molten In. The vapor pressure of In is negligible (~ 10^{-15} bars) at the synthesis temperature.

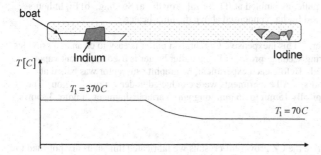

Figure 1. Synthesis of InI in a two-zone furnace.

Crystal growth experiments

Our initial attempts in 2008 to pull InI crystals in a high pressure CZ puller have revealed a significant evaporation from the melt, which could not be noticeably reduced using high pressure argon. The high pressure puller provided poor visibility (the melt at 360 C does not emit light) and was difficult to clean. Therefore, we have designed an easy to clean 1 bar argon pressure Czochralski puller, which provides an excellent visibility (figure 2). The puller is comprised of a 5 cm diameter silica tube, connected to a crystal withdrawal and rotation mechanism (borrowed from an Arthur D. Little (ADL) MP furnace) using standard 2 " QF flanges. The hot zone consists of a graphite susceptor heated by a 1.5 MHz induction power

Figure 2. Czochralski puller assembled at IIT for InI growth. a) Seeding; b) Eight hours after seeding, a layer of red InI has condensed above the after heater.

supply, and an after heater. This inexpensive Czochralski puller is easy to clean and provides an excellent visibility during seeding, provided that an after heater is used to prevent vapor deposition ot the silica wall. Before each experiment, he graphite susceptor was baked at ~ 600^0C in vacuum for 3 hours. All experiments were conducted under 1 atm of Argon. The obtained crystals were typically 15mm to 26 mm long with variable diameter. Figure 3 shows three such crystals.

Characterization

In order to characterize the CZ grown InI crystals we fabricated thin slabs dry polished to 5 μm finish then etched 1 minute in Br-MeOH solution (3%). Palladium contacts about 10/20 nm thick with diameters of 1.2 mm (bottom) and 3.5 mm (top)

Figure 3. Czochralski grown InI crystals

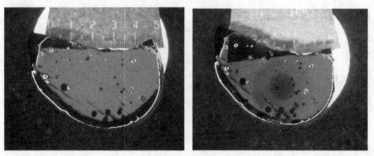

Figure 4. InI sample grown at IIT by Czochralski method, 6x4x0.77 mm³. Image on left is taken by transmission before Pd contact deposition. Image on right shows the sample after contact deposition.

On sample shown in figure 4, we measured the current-voltage characteristic and found the bulk resistivity is relatively low for a wide bandgap semiconductor, $1.6 \times 10^8 \Omega$cm, due probably to a high density of growth defects (figure 5).

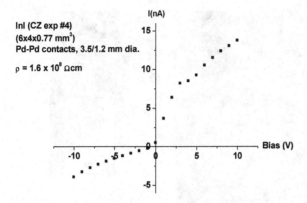

Figure 5. Current-Voltage characteristic of CZ InI sample. Bulk resistivity is $1.6 \times 10^8 \Omega$cm

The semiconductor properties of CZ grown InI were tested using both a phoconductivity method and also the radiation detection response to both alpha particles and gamma-rays photons. Figure 6 reveals that holes are the main charge carriers and the bandgap is 2.0 eV

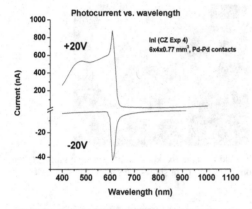

Fig. 6. Photocurrent as function of wavelength. Thin Pd contacts successively biased as anode then cathode (at 20 V) was illuminated with light of from 400 to 1000 nm wavelength. Photopeaks at 615 nm indicate that the bandgap is 2.0 eV and holes are the main charge carriers.

The photocurrent under constant above the bandgap illumination of the semitransparent anode indicate a value for $\mu\tau_h > 1.2 \times 10^{-3}$ cm^2/V as can be easily from graph in Fig. 7.

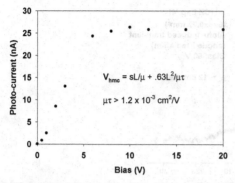

Figure 7. Photocurrent as function of bias. The semitransparent anode was illuminated with low intensity light of 500 nm.

The $\mu\tau$ product for holes, measured using the detector response to alpha particles shown in figure 8 differs substantially from the value obtained from DC photocurrent measurement under constant illumination due to the different nature of the two experiments: one is dealing with charge in the pulse mode of operation, the other in quasi-continuous mode when much longer de-traping time leads to a much higher value.

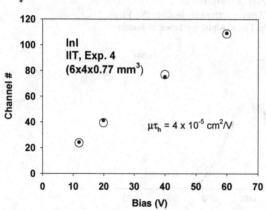

Figure 8 InI detector, holes mobility-lifetime product using alpha particles measurement. Hecht fit of pulse height alpha peaks as function of bias. $\mu\tau_h$ is 4×10^{-5} cm^2/V.

The hole mobility is roughly estimated to 12 cm^2/Vs from alpha particle induced transient timing as can be seen in figure 9.

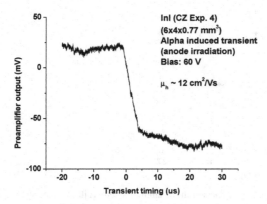

Figure 9 Estimate of hole mobility from alpha induced transient timing. Calculation is done using the transit time of the fast component of the transient.

Finally the InI crystal grown at IIT by Czochralski method was tested for detector response to alpha particles and gamma rays as well. The results are presented in the figures 10 and 11 respectively. The results are poor, justifiable by low $\mu\tau$ value and high density of defects.

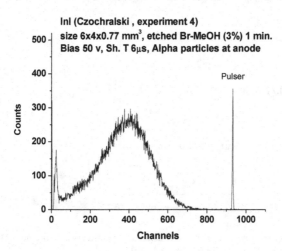

Figure 10. InI detector response to alpha particles (Am-241)

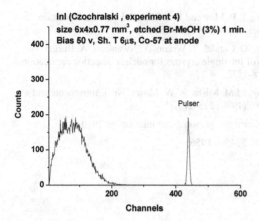

InI (Czochralski , experiment 4)
size 6x4x0.77 mm³, etched Br-MeOH (3%) 1 min.
Bias 50 v, Sh. T 6µs, Co-57 at anode

Figure 11. Co-57 spectrum obtained with a InI sample with size 6x4x0.77 mm³ at a bias of 50V and shaping time of 6 µs.

CONCLUSIONS

Growth of InI crystals using the Czochralski method was demonstrated for the first time. The crystals showed good sensitivity to light, alpha particles and gamma rays, but need improvement in crystalline and chemical perfection. The next step needed is to purify the starting InI by several steps of sublimation and recondensation.

ACKNOWLEDGMENTS

The authors are indebted to the National Nuclear Security Agency of the US Department of Energy for its support of this work under the contract DE-PS52-09NA293.

REFERENCES

1. H.M. Hobgooda, B.W. Swansona and R.N. Thomas, "Czochralski growth of CdTe and CdMnTe from liquid encapsulated melts", J. Crystal Growth 85 , (1987) 510-520

2. G.W. Blackmore, S.J. Courtney, A. Royle, N. Shaw, A.W. Vere,"Boron-segregation in Czochralski-grown CdTe" J. Crystal Growth 85 (1987) 335-340

3. I. Barin, "Thermophisical Data of Pure Substances", 2nd Edition, VCH Verlagsgesellschaft, Weinheim, Germany, 1993.

4. R.W.G. Wyckoff, "Crystal Structures" 2nd Edition, Volume 1, Interscience, New York, 1963.

5.   M.R. Squillante, C. Zhou, J. Zhang, L.P. Moy and K.S. Shah, InI Nuclear Radiation Detectors IEEE Trans. on Nucl. Sci. 40/ 4 (1993) 364-366.

6.   P. Bhattacharya, M. Groza, Y. Cui, D. Caudel, T. Wrenn, A. Nwankwo, A. Burger, G. Slack, A.G. Ostrogorsky, Growth of InI single crystals for nuclear detection applications', J. Crystal Growth 312 (2010) 1228-1232.

7.   K.S. Shah, P. Bennett,   L. P. Moy, M.M. Misra, W.W. Moses, Nucl. Intruments and Methods in Pyisics Research A 380 (1996) 215-219.

8.   A. Croll, University of Freiburg, Germany, personal communication 2010.

9.   E.A. Peretti, J. Am. Chem. Soc. 78, 5745-6(1956).

Mater. Res. Soc. Symp. Proc. Vol. 1341 © 2011 Materials Research Society
DOI: 10.1557/opl.2011.1484

Dysprosium-Containing Nanocrystals for Thermal Neutron Detection

Antonio C. Rivera[1], Natasha N. Glazener[1], Nathaniel C. Cook[1], Nathan J. Withers[1], John B. Plumley[1], Brian A. Akins[1], Ken Carpenter[2], Gennady A. Smolyakov[1], Robert D. Busch[2], and Marek Osiński[1]

[1]Center for High Technology Materials, University of New Mexico, 1313 Goddard SE, Albuquerque, NM 87106-4343
Tel. (505) 272-7812; Fax (505) 272-7801; E-mail: osinski@chtm.unm.edu

[2]Department of Chemical and Nuclear Engineering, 1 University of New Mexico, Albuquerque, NM 87131

ABSTRACT

We propose a novel concept of optical detection of thermal neutrons in a passive device that exploits transmutation of Dy-164, a dominant, naturally occurring isotope of dysprosium, into a stable isotope of either holmium Ho-165 or erbium Er-166. Combination of the high thermal neutron capture cross section of ~2,650 barns and transmutation into two other lanthanides makes Dy-164 a very attractive alternative to traditional methods of neutron detection that will be completely insensitive to gamma irradiation, thus reducing greatly the likelihood of false alarms. The optically enabled neutron detection relies on significant differences in optical properties of Dy, Ho, and Er that are not sensitive to a particular isotope, but change considerably from one element to another. While the concept applies equally well to bulk materials and to nanocrystals, nanocrystalline approach is much more attractive due to its significantly lower cost, relative ease of colloidal synthesis of high quality nanocrystals (NCs), and superior optical and mechanical properties of NCs compared to their bulk counterparts. We report on colloidal synthesis of DyF_3 NCs, both doped and undoped with Ho and co-doped with Ce and Eu to enhance their optical properties. We also report on DyF_3:10%Ce and DyF_3:10%Eu NCs irradiated with thermal neutrons from a Cf-252 source and their optical characterization.

INTRODUCTION

Standard detectors of slow neutrons rely on the $^{10}B(n,\alpha)$, $^{6}Li(n,\alpha)$, or $^{3}He(n,p)$ reactions [1]. The thermal neutron cross section for the $^{10}B(n,\alpha)$ reaction is 3840 barns, and the natural abundance of ^{10}B is 19.8%. The most common detector based on the boron reaction is a BF_3 gas tube. Boron-loaded scintillators are also used, although they encounter the challenge of discriminating between gamma rays backgrounds and gamma rays due to neutrons. The thermal neutron cross section for the $^{6}Li(n,\alpha)$ reaction is 940 barns, and the natural abundance of ^{6}Li is only 7.4%. This requires enrichment of ^{6}Li isotope. Moreover, detectors using ^{6}Li suffer from their sensitivity to gamma radiation. The thermal neutron cross section for the $^{3}He(n,p)$ reaction is 5330 barns, but its natural abundance of only 0.0001% results in a very high cost ($1,000 per liter for US commercial use [2]). Currently, all ^{3}He production comes solely from the refurbishment and dismantlement of the nuclear stockpile as a byproduct from the radioactive decay of tritium, where it is separated during the tritium cleaning process. A further problem with all of these neutron detection methods is the need for active electronics to detect the particle emitted from the nucleus that has absorbed a neutron, increasing size, cost, and danger of compromising the mission if used for clandestine activities.

We propose a novel concept of optical detection of thermal neutrons that exploits transmutation of a dominant naturally occurring isotope of dysprosium [164]Dy (28.22% natural abundance [3]) into a stable isotope of either holmium [165]Ho or erbium [166]Er, as illustrated in Fig. 1. The high neutron capture cross section (~2650 barns) combined with high natural abundance (28.2%) of [164]Dy make it a very attractive candidate for low-cost neutron detection. As result of neutron capture, [164]Dy forms a metastable isomer [165m]Dy, which can either relax to [165]Dy, or, upon the capture of a second neutron (with the cross section of 2000 barns), can be converted into [166]Dy. [165]Dy itself has two decay routes, first with the half-life of 140 min to a stable isotope of holmium [165]Ho, or upon the capture of a second neutron (with the cross section of 3530 barns) to [166]Dy. The latter has a half-life of 81.6 hours, producing a radioactive daughter [166]Ho, which then decays with the half-life of 26.8 hours into a stable isotope of erbium [166]Er.

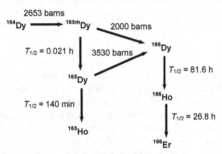

Figure 1. Decay path and lifetimes for [165m]Dy after neutron interaction (after Ref. 4).

The key concept in how to use these processes for optically-enabled neutron detection relies on significant differences in optical properties of Dy, Ho, and Er that are not sensitive to a particular isotope, but change considerably from one element to another.

Thin foils of metallic dysprosium have been in use for slow neutron radiography for a long time [5], where beta particle emission was used as means of quantifying the exposure. In our concept, instead of observing the decay events, the same information can be retrieved by optical interrogation of the transmuted elements. This allows a completely passive detector that can be examined days or even months after exposure for the presence of neutrons. It should be emphasized that for low doses, multiple readings and subsequent exposures are possible on the same sample, as long as the results of previous exposures are accounted for.

While the concept described above applies equally well to bulk materials and to nanocrystals (NCs), we believe that nanocrystalline approach is much more attractive due to its significantly lower cost, relative ease of colloidal synthesis of high quality NCs with controlled composition, and superior optical and mechanical properties of NCs compared to their bulk counterparts. One particular advantage of NCs for neutron detection is that they can be integrated into a transparent polymer host without causing optical scattering, and the host can serve the dual functions of making the neutron detector mechanically robust as well as moderating incoming neutrons, improving the probability of interaction and detection sensitivity.

Here, we report on colloidal synthesis of DyF$_3$ NCs both doped and undoped with Ho, and co-doped with Ce and Eu to enhance their optical properties. We also report on DyF$_3$:10%Ce and DyF$_3$:10%Eu NCs irradiated with thermal neutrons from a [252]Cf source, and their optical characterization.

EXPERIMENT

Synthesis of DyF₃ NCs Capped with Oleic Acid

The synthesis of DyF_3 NCs capped with oleic acid was modified after [6]. The following chemical reaction is expected to occur at the nucleation phase of oleic-acid-capped DyF_3 NCs:

$$CH_3OH + 1 \text{ mmol } Dy(NO_3)_3 \cdot H_2O + 3.0 \text{ mmol } NH_4F \rightarrow 3 \text{ mmol } NH_4NO_3 + 1 \text{ mmol } DyF_3/OH$$

In order to keep the reaction under argon, a three-neck borosilicate glass was connected to a Schlenk line gas manifold. 35 mL of 1.6M NH_4F was added to the three-neck flask and brought to 60 °C. Then 2 mL of 1.5M $Dy(NO_3)_3 \cdot H_2O$ was added to the three neck flask and allowed to react for 2 hours. When the solution began to cool, 0.5 mL of oleic acid was added to coat the NCs. After allowing the mixture to cool to room temperature, it was centrifuged at 4000 rpm and a white precipitate was collected in toluene. Doped samples of DyF_3:10%Ce, DyF_3:10%Eu, DyF_3:5%Ce,5%Eu, DyF_3:10%Ho, DyF_3:10%Ce,1%Ho, and DyF_3:10%Eu,1%Ho NCs were also prepared by reducing the amounts of the Dy precursor and adding proportional amounts of the dopant precursors. For comparison, we have also prepared HoF_3:10%Ce and HoF_3:10%Eu NCs by replacing the nitrate precursors in the appropriate proportions. By visual inspection, the HoF_3-based NCs were pink, while the DyF_3-based NCs were white.

Structural Characterization of Dy- and Ho-Containing NCs Capped with Oleic Acid

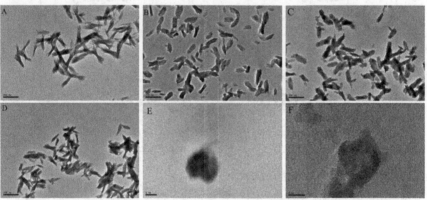

Figure 2. TEM images of A) DyF_3; B) DyF_3:10%Ce; C) DyF_3:10%Eu; D) DyF_3:5%Ce,5%Eu; E) HoF_3:10%Ce; F) HoF_3:10%Eu. Scale bars are 200 nm on images A-D, and 5 nm on images E, F.

Fig. 2 shows bright field images of the DyF_3, DyF_3:10%Ce, DyF_3:10%Eu, DyF_3:5%Eu,5%Ce, HoF_3:10%Ce, and HoF_3:10%Eu NCs, taken on JEOL 2010 TEM. According to the TEM data, the DyF_3-based NCs are rods that have a length varying from 50 nm to 200 nm and are around 10 – 50 nm in width, which agrees with the previous work by Hoffmann reporting on making DyF_3 nanofibers [7]. As shown in the HRTEM images E, F, the Ho-containing NCs tend to form platelets around 10 – 20 nm in diameter. The difference in shapes is caused by different thermodynamic growth preferences between DyF_3 and HoF_3 [7].

Fig. 3 shows a high resolution image of DyF_3:10%Ce NC sample that displays fringes. Using an FFT image we can see that the crystal is being viewed along the [0001] zone axis and matches the hexagonal crystal planes expected in DyF_3.

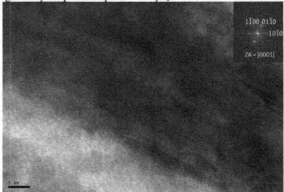

Figure 3. High-resolution TEM image of DyF_3 hcp lattice fringes and zone axis data. Scale bar: 5 nm.

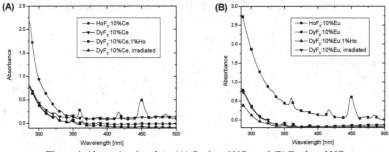

Figure 4. Absorption data from (A) Ce-doped NCs, and (B) Eu-doped NCs.

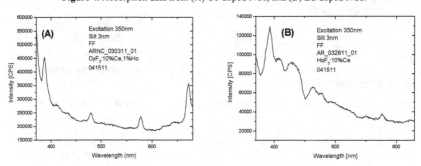

Figure 5. (A) Photoluminescence of DyF_3:10%Ce NC sample excited at 350 nm; (B) Photoluminescence of HoF_3:10%Ce NC sample excited at 350 nm.

Optical Characterization of Dy- and Ho-Containing NCs Capped with Oleic Acid

Fig. 4 presents the absorption spectra data of various Dy- and Ho-containing NC samples comparing the doped, co-doped, and irradiated samples. Distinctive peaks for the Ho-containing NCs can be seen at 450 nm, 417 nm, and 360 nm. These lines are present only in the HoF_3-based samples and are not observed in the DyF_3-based samples.

We measured photoluminescence (PL) of the DyF_3:10%Ce and HoF_3:10%Ce NCs (Fig. 5) and observed definite peaks at 425 nm and 463 nm from Ho that were not present in the DyF_3:10%Ce sample. We also observed these peaks in the Eu-doped samples (Fig. 6).

These observations led us to attempt to synthesize DyF_3:10%Ce,1%Ho and DyF_3:10%Eu,1%Ho NCs and to irradiate part of both DyF_3-based samples in order to determine if we could see the peaks, observed in the doped HoF_3 samples, begin to develop.

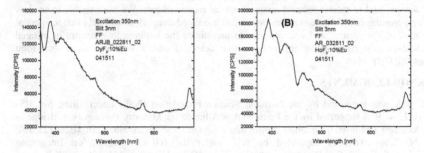

Figure 6: (A) Photoluminescence of DyF_3:10%Eu NC sample excited at 350 nm; (B) Photoluminescence of HoF_3:10%Eu NC sample excited at 350 nm.

Neutron Irradiation Experiments

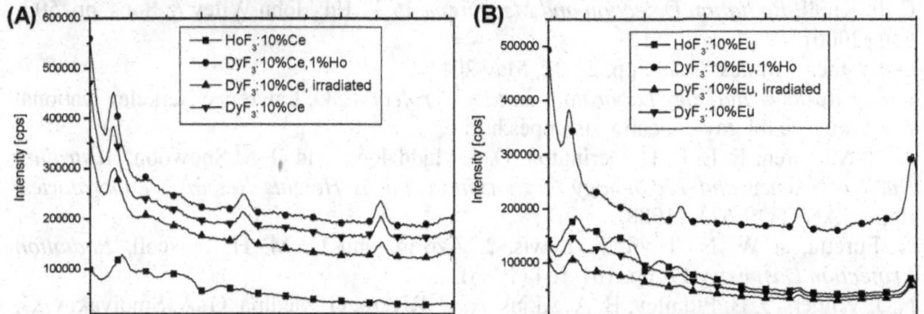

Figure 7: (A) Ce-doped NC samples excited at 350 nm; (B) Eu-doped NC samples excited at 350 nm.

For the neutron irradiation experiment, samples of DyF_3:10%Ce and DyF_3:10%Eu NCs were placed 12.7 cm from a ^{252}Cf source with an activity of 1.17 mCi, emitting 5.3×10^6 neutrons/s and located inside of a 10-inch diameter polyethylene sphere. The flux through the sample was 2,609 neutrons/cm^{2}s. The samples were left in this configuration for 3.919×10^5 s, giving them a total fluence of 1.023×10^9 neutrons/cm^2.

The photoluminescence measurements were repeated for the irradiated samples (Fig. 7) and the results were compared with the data from the non-irradiated samples, with the purpose of identifying any holmium-related features in the spectra. In both the Ce- and Eu-doped samples we were unable to see any measurable changes in the visible range to indicate whether Ho was present.

CONCLUSIONS

We have shown the ability to optically distinguish DyF$_3$:10%Ce NC samples from HoF$_3$:10%Ce NC samples. Our next step is to find a combination of dopants and co-dopants to enhance this shift to achieve maximum spectral differentiation between the non-irradiated and irradiated DyF$_3$ NC samples, and to extend spectral range of optical measurements. We were unable to see low doping percentages of holmium in our doped and irradiated samples, most likely due to spectral limitation of our system, and will continue to investigate the luminescence spectra in infrared and mid-infrared range using infrared-sensitive detectors and a separate Fourier-transform infrared (FTIR) setup.

ACKNOWLEDGMENTS

This work was supported by the National Science Foundation (NSF) under Grant No. IIS-1016352. ACR is supported by the Louis Stokes Alliance for Minority Participation, Bridge to the Doctorate (AMP-BD) Graduate Fellowship Program (NSF Grant No. EHR-1026412). JBP and NJW are Trainees supported by NSF under the IGERT program on Integrating Nanotechnology with Cell Biology and Neuroscience (Grant No. DGE-0549500). NJW is also a recipient of DTRA Student Research Associateship (Grant No. DTRA-01-03-D-0009-0022).

REFERENCES

1. G. F. Knoll, *Radiation Detection and Measurement*, 3rd Ed., John Wiley & Sons, pp. 507-520 (2000).
2. D. Kramer, *Physics Today*, pp. 20-22, May 2011.
3. R. B. Firestone, *Berkeley Laboratory Isotopes Project*, E. O. Lawrence Berkeley National Lab., http://ie.lbl.gov/education/isotopes.htm
4. A. B. McLaren, E. L. R. Hetherington, D. J. Maddalena, and G. M.Snowdon, *Australian Nuclear Science and Technology Organization, Lucas Heights Research Laboratories*, Report ANSTO/E673, (1988).
5. C. Furetta, J. W. N. Tuyn, F. Lewis, J. Azorin, and C. M. H. Driscoll, *Radiation Protection Dosimetry* **17**, pp. 161-164 (1986).
6. N. J. Withers, J. B. Plumley, B. A. Akins, A. C. Rivera, G. Medina, G. A. Smolyakov, G. S. Timmins, and M. Osiński, *Colloidal Quantum Dots for Biomedical Applications V* (M. Osiński, W. J. Parak, T. M. Jovin, and K. Yamamoto, Eds.), San Francisco, CA, 23-25 Jan. 2010, *Proc. SPIE* **7575**, Paper 75750Z (10 pp.).
7. M. M. Hoffmann, J. S. Young, and J. L. Fulton, *J. Nuclear Science*, **35** (16) 4177-4183 (2000).

Mater. Res. Soc. Symp. Proc. Vol. 1341 © 2011 Materials Research Society
DOI: 10.1557/opl.2011.1112

Semiconductor Detectors Fabricated from TlBr Crystals

Keitaro Hitomi[1], Tsutomu Tada[1], Seong-Yun Kim[1], Yan Wu[1], Hiromichi Yamazaki[1],
Tadayoshi Shoji[2] and Keizo Ishii[3]
[1]Cyclotron and Radioisotope Center, Tohoku University,
Sendai 980-8578, Japan
[2]Department of Electronics and Intelligent Systems, Tohoku Institute of Technology,
Sendai 982-8577, Japan
[3]Department of Quantum Science and Energy Engineering, Graduate School of Engineering,
Tohoku University,
Sendai 980-8579, Japan

ABSTRACT

Frisch collar detectors were fabricated from TlBr crystals with the dimensions of 2 mm ×
2 mm × 4.4 mm. Spectroscopic performance of the TlBr Frisch collar detectors was evaluated at
room temperature. An energy resolution of 2.9% FWHM at 662 keV was obtained from the
detector without the depth correction. The detector exhibited stable spectral performance for 12
hours. Direct measurements of electron mobility-lifetime products were performed with the
detectors. The TlBr crystals exhibited the electron mobility-lifetime products of $\sim 10^{-3}$ cm^2/V at
room temperature.

INTRODUCTION

Thallium bromide (TlBr) is a compound semiconductor promising for fabrication of
efficient radiation detectors with high energy resolutions at room temperature. The TlBr crystals
exhibit high photon stopping power originating from its high atomic numbers (Tl: 81, Br: 35)
and high density (7.56 g/cm^3). The wide bandgap energy (2.68 eV) results in the high resistivity
of TlBr crystals ($\sim 10^{10}$ Ω·cm) at room temperature [1], enabling the device to operate with low
noise. The charge transport properties in TlBr crystals were improved significantly by
purification of the starting material [2]. The mobility-lifetime products for electrons and holes in
TlBr crystals were reported to be $\sim 10^{-3}$ cm^2/V and $\sim 10^{-4}$ cm^2/V, respectively [2]. The TlBr melts
congruently at the melting point of 460 °C and has no destructive phase transition between the
solidification and room temperature. The purification and crystal growth of TlBr can be
performed simply from the melt. Due to its attractive physical properties, semiconductor
detectors fabricated from TlBr crystals were studied by numerous researchers [1-8].

The thickness of TlBr detectors with planar electrodes was limited to less than 1 mm due
to the hole trapping effects. Pixelated TlBr detectors were fabricated for overcoming the problem
by the small pixel effect [9] and demonstrated improved energy resolutions [8,10].

In order to solve the hole trapping problem, the Frisch collar detectors were fabricated
from TlBr crystals in this study. The Frisch collar detector operates as a single polarity charge
sensing device, in which the induced charge on the anode depends mainly on the collection of
electrons and scarcely affected by the hole movement. The Frisch collar detector design was
applied to CdZnTe, HgI$_2$ and TlBr, and the detectors exhibited excellent spectroscopic
performance [11-15].

In this study, spectroscopic performance of TlBr Frisch collar detectors was evaluated at room temperature. Direct measurements of electron mobility-lifetime products $\mu_e\tau_e$ in TlBr crystals were performed at room temperature. The results of the detector evaluations are presented in this paper.

EXPERIMENT

Crystal growth and detector fabrication

Commercially available TlBr powder with purity of 99.999% was employed for the crystal growth. Further purification of the starting material was performed by the zone refining method. After the purification, TlBr crystals were grown by the traveling molten zone method. The detailed crystal growth processes were described elsewhere [2].

Frisch collar detectors were fabricated from the grown crystals. Bar-shaped crystals with the dimensions of 2 mm × 2 mm × 4.4 mm were cut from the grown TlBr ingot using a diamond wire saw. The 2 mm × 2 mm surfaces were polished mechanically in order to remove the damaged surface layers originating from the cutting process. Planar electrodes were formed on the 2 mm × 2 mm surfaces by vacuum evaporation of Tl. Applying Tl electrodes to TlBr detectors is effective for suppressing the polarization phenomena at room temperature [16,17]. Since Tl is susceptible to oxidation, Al was evaporated onto the Tl electrodes to form an overcoat. The side surfaces of the bar-shaped crystal were wrapped with Teflon tape. A Frisch collar electrode was constructed by wrapping Al foil around the side surfaces. The Frisch collar covered the entire side surfaces of the device. Figure 1 (a) shows the fabricated TlBr planar detector and TlBr Frisch collar detector. The device was mounted on a substrate and thin Pd wires were attached to the electrodes (cathode, anode and Frisch collar) with conductive adhesive as shown in figure 1 (b).

Figure 1. (a) A TlBr planar detector (left) and a TlBr Frisch collar detector (right); (b) A TlBr Frisch collar detector mounted on a substrate. The dimensions of the crystals were 2 mm × 2 mm × 4.4 mm.

Device testing

Spectroscopic performance of the TlBr Frisch collar detectors was evaluated at room temperature. The TlBr Frisch collar detector was placed in a shield box. The cathode and the anode were connected to charge-sensitive preamplifiers (CLEAR PULSE 580 K). The Frisch collar electrode and the cathode were maintained at ground potential. Positive bias voltage was applied to the anode. The cathode was irradiated with a ^{137}Cs calibration source. The output signals from the preamplifiers were fed into shaping amplifiers (CANBERRA 2025 for the cathode and ORTEC 673 for the anode). The shaping times for the cathode and the anode were 6 μs. The peak amplitude for the cathode and anode pulses was analyzed by a multi-channel ADC board. The data were analyzed on a PC event by event to obtain pulse high spectra as a function of the cathode to anode signal ratio. The gamma-ray interaction depth is determined by taking the cathode to anode signal ratio in single polarity charge sensing devices [18]. The cathode to anode signal ratios were grouped into 21 bins in the measurements.

Direct measurements of electron mobility-lifetime products $\mu_e\tau_e$ in the TlBr crystals were performed at room temperature. Pulse height spectra of ^{137}Cs for the near cathode events were acquired with the anode of the TlBr Frisch collar detector at different bias voltages. The $\mu_e\tau_e$ value is derived from the equation [19]:

$$\mu_e\tau_e = \frac{D^2}{\ln(N_1/N_2)}\left(\frac{1}{V_2} - \frac{1}{V_1}\right),\tag{1}$$

where N_1 and N_2 are the photopeak centroids of the spectra recorded for the near cathode events at bias voltages of V_1 and V_2 across the detector thickness D, respectively.

DISCUSSION

Spectroscopic performance

Figure 2 shows ^{137}Cs spectra obtained from a 2 mm × 2 mm × 4.4 mm TlBr Frisch collar detector. The anode bias voltage was 500 V. The acquisition time was 2 hours in real time. The anode spectrum exhibited an energy resolution of 2.9% FWHM at 662 keV. The Tl X-ray escape peak and the backscatter peak are discernible in the spectrum. The K X-rays of Ba from the ^{137}Cs source were detected by the detector, which confirmed that the electrons created by the Ba K X-rays near the cathode surface traversed the entire detector length, implying less electron trapping in the crystal. The anode pulse height exhibited no significant depth (the cathode to anode signal ratio) dependence since the hole trapping effects were almost eliminated by the Frisch collar. On the other hand, the cathode spectrum was distorted due to the incomplete collection of holes. The cathode pulse height exhibited significant depth dependence, which implied that the Frisch collar caused no marked effect for the cathode signal. Slight depth dependence of the anode pulse height because of the electron trapping and the variation of the weighting potential was corrected by aligning the photopeak centroids to construct the depth corrected spectrum as shown in figure 3. Improvement of the energy resolution from 2.9% FWHM to 2.5% FWHM at 662 keV was achieved by the depth correction. Long-term stability of the TlBr Frisch collar detector was evaluated by acquiring ^{137}Cs spectra continuously for 12 hours at room temperature. Figure 4 shows ^{137}Cs spectra obtained from the detector as a function of the elapsed time. The bias voltage of 500 V was applied to the detector. Each spectrum was acquired for 1 hour. The detector exhibited stable spectral performance for 12 hours.

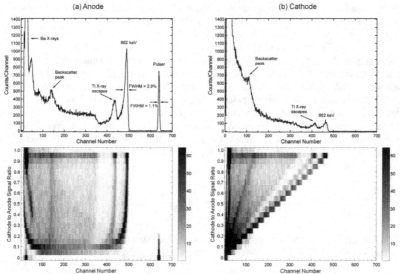

Figure 2. (a) ^{137}Cs spectrum obtained from the anode of a 2 mm × 2 mm × 4.4 mm TlBr Frisch collar detector (top) and the relationship between the anode pulse height and the cathode to anode signal ratio (bottom); (b) ^{137}Cs spectrum obtained from the cathode of the TlBr Frisch collar detector (top) and the relationship between the cathode pulse height and the cathode to anode signal ratio (bottom). The measurements were performed at room temperature.

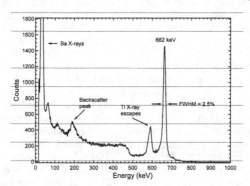

Figure 3. Depth-corrected ^{137}Cs spectrum obtained from the anode of the 2 mm × 2 mm × 4.4 mm TlBr Frisch collar detector. The measurement was performed at room temperature.

114

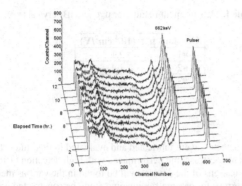

Figure 4. ^{137}Cs spectra obtained from the 2 mm × 2 mm × 4.4 mm TlBr Frisch collar detector as a function of the elapsed time after the bias voltage application. The bias voltage of 500 V was applied to the detector irradiated with the ^{137}Cs source constantly during the measurements for 12 hours.

Direct measurements of mobility-lifetime products

Direct measurements of electron mobility-lifetime products were performed for three 2 mm × 2 mm × 4.4 mm TlBr crystals obtained from the same ingot. Figure 5 shows pulse height spectra of ^{137}Cs for the near cathode events acquired with the anode of the TlBr Frisch collar detector at different bias voltages. The photopeak centroid was shifted from 470 channel to 492 channel when the bias voltage was increased from 400 V to 500 V. Using the equation (1), the $\mu_e\tau_e$ value was estimated to be 2.1×10^{-3} cm^2/V. The measurement results for the $\mu_e\tau_e$ values in the three TlBr crystals are summarized in Table I. The three TlBr crystals exhibited similar $\mu_e\tau_e$ values. The estimated $\mu_e\tau_e$ values are well consistent with the reported data [10].

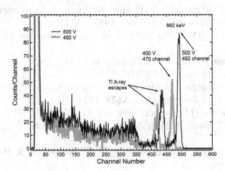

Figure 5. ^{137}Cs spectra for the near cathode events (the cathode to anode signal ratio ≥ 0.95) acquired with the anode of the 2 mm × 2 mm × 4.4 mm TlBr Frisch collar detector at different bias voltages, 400 V and 500 V. The measurements were performed at room temperature.

115

Table I. Measurement results for $\mu_e\tau_e$ in TlBr crystals.

Crystal	$\mu_e\tau_e$ (10^{-3} cm^2/V)
#1	2.1
#2	1.4
#3	1.7

CONCLUSIONS

TlBr Frisch collar detectors were fabricated and evaluated in this study. The TlBr detectors exhibited excellent spectroscopic performance with high detection efficiency at room temperature. Possible applications of the devices will be found in the various medical and industrial fields. Detector arrays for gamma-ray imaging could be constructed with the TlBr Frisch collar detectors. Direct measurements of electron mobility-lifetime products were performed with the detectors. The value of mobility-lifetime products for electrons in TlBr crystals approaches the value reported for CdTe and CZT crystals. Future research efforts will be directed toward improving the crystal growth and detector fabrication processes for TlBr detectors, evaluating the timing resolutions of the devices and establishing the methods for mass production of TlBr detectors.

REFERENCES

1. K. Hitomi, T. Murayama, T. Shoji, T. Suehiro, and Y. Hiratate, *Nucl. Instrum. Meth.* **A428**, 372 (1999).
2. K. Hitomi, T. Onodera, and T. Shoji, *Nucl. Instrum. Meth.* **A579**, 153 (2007).
3. K.S. Shah, J.C. Lund, F. Olschner, L. Moy, and M.R. Squillante, *IEEE Trans. Nucl. Sci.* **36**, 199 (1989).
4. V. Gostilo, A. Owens, M. Bavdaz, I. Lisjutin, A. Peacock, H. Sipila, and S. Zatoloka, *Nucl. Instrum. Meth.* **A509**, 47 (2003).
5. A. Owens, M. Bavdaz, G. Brammertz, V. Gostilo, N. Haack, A. Kozorezov, I. Lisjutin, A. Peacock, and S. Zatoloka, *Nucl. Instrum. Meth.* **A497**, 359 (2003).
6. V. Kozlov, M. Kemell, M. Vehkamäki, and M. Leskelä, *Nucl. Instrum. Meth.* **A576**, 10 (2007).
7. A.V. Churilov, G. Ciampi, H. Kim, L.J. Cirignano, W.M. Higgins, F. Olschner, and K.S. Shah, *IEEE Trans. Nucl. Sci.* **56**, 1875 (2009).
8. H. Kim, A. Churilov, G. Ciampi, L. Cirignano, W. Higgins, S. Kim, P. O'Dougherty, F. Olschner, and K. Shah, *Nucl. Instrum. Meth.* **A629**, 192 (2011).
9. H.H. Barrett, J.D. Eskin, and H.B. Barber, *Phys. Rev. Lett.* **75**, 156 (1995).
10. K. Hitomi, T. Onodera, T. Shoji, Y. Hiratate, and Z. He, *IEEE Trans. Nucl. Sci.* **55**, 1781 (2008).
11. W. J. McNeil, D. S. McGregor, A. E. Bolotnikov, G. W. Wright, and R. B. James, *Appl. Phys. Lett.* **84**, 1988 (2004).
12. A. Kargar, A. M. Jones, W. J. McNeil, M. J. Harrison, and D. S. McGregor, *Nucl. Instrum. Meth.* **A558**, 497 (2006).

13. A. E. Bolotnikov, J. Baker, R. DeVito, J. Sandoval, and L. Szurbart, *IEEE Trans. Nucl. Sci.* **52**, 468 (2005).
14. E. Ariesanti, A. Kargar, and D.S. McGregor, *Nucl. Instrum. Meth.* **A624**, 656 (2010).
15. H. Kim, L. Cirignano, A. Churilov, G. Ciampi, A. Kargar, W. Higgins, P. O'Dougherty, S. Kim, M. R. Squillante, and K. Shah, *Proc. SPIE* **7806**, 780604 (2010).
16. K. Hitomi, T. Shoji, and Y. Niizeki, *Nucl. Instrum. Meth.* **A585**, 102 (2008).
17. K. Hitomi, Y. Kikuchi, T. Shoji, and K. Ishii, *IEEE Trans. Nucl. Sci.* **56**, 1859 (2009).
18. Z. He, G.F. Knoll, D.K. Wehe, R. Rojeski, C.H. Mastrangelo, M. Hammig, C. Barrett, and A. Uritani, *Nucl. Instrum. Meth.* **A380**, 228 (1996).
19. Z. He, G.F. Knoll, and D.K. Wehe, *J. Appl. Phys.* **84**, 5566 (1998).

Mater. Res. Soc. Symp. Proc. Vol. 1341 © 2011 Materials Research Society
DOI: 10.1557/opl.2011.1113

UV emitting single crystalline film scintillators
grown by LPE method: current status and perspective

Yuriy Zorenko[1,3]; Vitaliy Gorbenko[1]; Volodymyr Savchyn[1]; Taras Voznyak[1];
Miroslaw Batentschuk[2]; Albrecht Winnacker[2]; Qi Xia[2]; Christoph Brabec[2]
[1]Electronic Department of Ivan Franko National University of Lviv, 79017 Lviv, Ukraine;
[2]Department of Materials Science and Engineering, University of Erlangen-Nuremberg, 91058
Erlangen, Germany
[3]Institute of Physics, Kazimierz Wielki University in Bydgoszcz, 85-090 Bydgoszcz, Poland

ABSTRACT

This work is dedicated to the development of new type of UV phosphors based on single
crystalline films (SCF) of aluminum garnet compounds grown by liquid phase epitaxy (LPE). The
development of two types of UV emitting SCF scintillators is reported in this work: 1) Pr-doped
SCF of Y-Lu-Al garnet having the intensive Pr^{3+} f-d luminescence in the 300-400 nm spectral
ranges with a decay time of about 13-18 ns; 2) SCF of Y-Lu-Al-garnet doped with Sc^{3+}
isoelectronic impurity emitting in the 290-400 nm range due to the formation of the $Sc_{Y,Lu}$ and Sc_{Al}
centers with a luminescence decay time in the order of several hundred ns.

INTRODUCTION

This work is related to the development of new type of phosphors based on single
crystalline films (SCF) of oxide compounds (aluminum garnets and perovskites, sapphire,
silicates, tungstates, i.e.) grown by liquid phase epitaxy (LPE). The field of applications of such
phosphors is rapidly extending in last decade and includes now screens for visualization of x-ray
images, α– and β-scintillators, cathodoluminescent light sources, laser media as well as
converters of LED radiation [1, 2].

The main goal of our work is to report on the development of UV emitting SCF scintillating
screens based on aluminum garnet compounds. The shift of emission spectra of SCF to the UV
range with respect to recently developed YAG:Ce and LuAG:Ce SCF scintillastors with emission in
the visible (450-750 nm) range [1] in principle can result in an increase of the light yield (LY) and
energy resolution of scintillators. Specifically, using the UV-emitting SCF scintillating screens can
significantly improve the spatial resolution of detectors for visualization of X-ray images [3].
Development of the raster scanning optical microscope technique also requires different UV
cathodoluminescent light sources for the investigation of biological objects.

The Pr^{3+} and Sc^{3+} ions show intensive emission in the UV range in single crystal (SC) of Y-
Lu-Al garnets [4, 5]. The single crystalline films (SCF) of these compounds also are considered
now for application as cathodoluminescent (CL) screen and scintillating screens for high-
resolution 2D-imaging [6, 7].

Our work is dedicated to grown by the liquid phase epitaxy (LPE) method and detail studying
the luminescence and scintillation properties of Pr^{3+} and Sc^{3+} doped SCFs of $Y_3Al_5O_{12}$ (YAG) and
$Lu_3Al_5O_{12}$ (LuAG) garnets. The following main features of the Pr^{3+} and Sc^{3+} luminescence in these
SCF are considered in this work: (i) dependence of intensity of the Pr^{3+} and Sc^{3+} luminescence in Y-
Lu-Al garnets on the activator concentration; (ii) influence of flux contamination on the light yield
(LY) of Pr^{3+} and Sc^{3+} luminescence in these SCF with respect to their single crystal (SC) analogues
[4, 5] and Ce-doped SCF analogues [1] of the same content, emitting in the visible range.

2. GROWTH OF SCF SCINTILLATORS AND EXPERIMENTAL TECHNIQUE

The series of Pr^{3+} and Sc^{3+} doped YAG and LuAG SCF scintillators were grown in University of Lviv by the LPE methods onto YAG substrates with the melt-solution (MS) based on the traditional $PbO-B_2O_3$ fluxes at relatively low (950-1100°C) temperatures as compared to Czochralski-grown single crystal (SC) analogues (1970-2000 °C). For the growth of YAG:Pr SCF onto YAG substrates we also used the novel lead-free $BaO-BaF_2-B_2O_3$ flux (these samples will be indicated below as YAG:Pr (BaO) SCF). The thickness of the SCF phosphors was in the 5-45 micron range. The growth rate for SCF grown from PbO flux was in 0.2-1.3 micron/min range. But for SCF grown from BaO based flux the growth rate was significantly lower (0.04-0.01).

Lower growth temperature of SCFs of garnets results in the absence of the Y_{Al} and Lu_{Al} antisite defects (AD) and decrease of the content of other type of defects in SCF in comparison with bulk SC analogues [1]. On the other hand, flux components can be introduced in the SCF and can influence their emission and scintillation properties. We have shown recently that the YAG:Ce and LuAG:Ce SCFs grown from the PbO-based flux usually contain lead ion contamination preferably in the Pb^{2+} charge state. It causes decrease of the LY and slowing-down the scintillation response [8]. Therefore, LY of UV emitting SCF phosphors can also strongly depend on the activator concentration and Pb^{2+} contamination level.

Due to significant difference in the segregation coefficient of the Sc (0.2-0.4) and Pr (0.04-0.1) ions in YAG and LuAG SCF grown on YAG substrates from PBO-based fluxes [9], the content of Sc_2O_3, and Pr_2O_3 oxides in MS was varied in the 1.25-6.7 and 3.1-8.6 mole % ranges, respectively. The concentration of these dopants in the SCF depends not only on the content of activated oxides in MS, but is also strongly influenced by the SCF growth temperature. Usually the activators/Pb ratio increase with increasing the growth temperature and vice versa [9]. Thus we have used relatively high temperatures above 950 °C for the SCF growth [9].

The content of Ce and Sc dopants in SCF was determined using a JEOL JXA-8612 MX electron microscope and is presented in Table 1 for different series of SCF phosphors.

The CL spectra of SCF phosphors were measured at 300 K with a set-up based on a DMR-4 monochromator and a FEU-106 PMT under pulsed electron-beam excitation (9 KeV, 100 μA) with pulse duration of 2 μs and a frequency of 3-30 Hz. The decay of luminescence was measured at 300 K in the time interval 0-200 ns under excitation by SR with pulse duration of 0.126 ns at Superlumi station (HASYLAB at DESY, Hamburg, Germany).

Relative LY of all the mentioned SCF scintillators was measured in comparison with that of the corresponding samples of YAG:Pr, LuAG:Pr, YAG:Sc, LuAG:Sc SCs and standard YAP:Ce SC using detector based on FEU-100 PMTs with maxima sensitivity ranges of 250-350 nm and a multi-channel single-photon counting system working in the time interval 0.5 micro-second under excitation by alpha-particles of Pu^{239} (5.15 MeV) sources. The penetration depth of alpha-particles in (Y-Lu)AG host is about 10-12 micron. The results of LY measurements of Pr^{3+} and Sc^{3+} doped (Y-Lu)AG SCF in comparison with their SC analogues and standard YAP:Ce SC sample are presented in Table 1.

3. LUMINESCENCE SPECTRA AND LIGHT YIELD OF SCF SCINTILLATORS

Pr^{3+}- doped garnets

The CL spectra of YAG:Pr and LuAG:Pr SCF showed intensive and fast emission in the 290-450 nm range with the main maxima at 323 and 305 nm, with the decay time of about 13 and 19 ns, respectively, caused by the $5d^1 4f^1 \rightarrow 4f^2 (^3H_4, {}^3H_5, {}^3H_6, {}^3F_{3(4)})$ transition of Pr^{3+} ions [4, 6] (Fig.1).

Table I. Relative LY of the best samples in series (Y-Lu)AG:Pr and (Y-Lu)AG:Sc SCF, grown by LPE methods in comparison with their SC analogues and YAP:Ce standard SC samples under excitation by α-particles of Pu^{239} sources (5.15 MeV) measured with the shaping time of 0.5 microsecond.

Scintillators	Activator content, at. %	Maximum of emission band, nm	Decay time of main emission component, ns	LY,%
YAG:Pr SCF	0.23	323	13	36.0
YAG:Pr (BaO) SCF	0.17	321	17	47
LuAG:Pr SCF	0.35	305	17	33.5
YAG:Pr SC	0.19	327	17.6	81.0
LuAG:Pr SC	0.31	308	18.6	80.0
YAG:Sc SCF	1.85	314	650	48.2
LuAG:Sc SCF	1.0	282	245; 390	46.2
LuAG:Sc SC	0.75-1.0	285	1330	60.0
YAG:Sc SC	0.39	314	580	56.8

The detailed consideration of luminescent properties of Pr^{3+} doped (Y-Lu)AG SCF is given in [6]. Here we mention only that the LY of (Y,Lu)AG:Pr SCFs is significantly lower (by 2.2-2.8 times) than that of their SC analogues (Table 1). The main reason for such low LY of the Pr^{3+}-doped SCF of garnets and perovskites is the strong quenching influence of the Pb^{2+} flux related impurity. From the results presented in Table 1, we conclude that significantly larger influence of Pb^{2+} ions on the UV emission of Pr^{3+} ions takes place in (Y,Lu)AG:Pr and (Y,Lu)AP:Pr SCFs in comparison with the influence of such impurity on the Ce^{3+} emission in the visible range in SCF of Ce-doped garnets [1]. The mechanism of the Pb^{2+}→Pr^{3+} energy transfer in Pr-doped SCF of garnets is considered in [10].

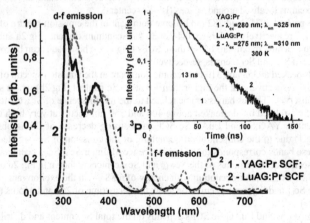

Figure 1. CL spectra and decay kinetics (insets) of Pr^{3+} luminescence in YAG:Pr (1a) and LuAG:Pr (2a) SCF at 300 K.

We observed also the acceleration of the decay kinetics of the Pr^{3+} luminescence in YAG:Pr and LuAG:Pr SCF in comparison with their SC analogues. The decay curve for both SCF is widely non-exponential and can be characterized by the mean decay time of 13 and 17 ns, respectively (Fig.1a, insert), which is different from the lifetime of 17.5 and 20 ns in SC of these garnets [4]. Such an shortening of the decay time of the Pr^{3+} luminescence in YAG:Pr and LuAG:Pr SCF is caused by the energy transfer from Pr^{3+} ions to Pb^{2+}-related centers [10].

Contrary to these SCFs, the Pr^{3+} emission decay in YAG:Pr (BaO) SCF is single-exponential with the decay time of 17 ns which is very close to the decay time of the Pr^{3+} emission in YAG:Pr SC (Table 1). The LY of YAG:Pr (BaO) SCF is also notable increase with respect of YAG:Pr (BaO) SCF up to 45 % of that for YAP:Ce SC (Table 1). That reflects the great application potential of BaO-based fluxes for growth of SCF phosphors. However, the main problem of using the BaO-based flux is its high viscosity and surface tension. Low SCF growth rate resulted in worse surface morphology and uniformity of SCF properties [11]. We also did not succeed with the hetero-epitaxial crystallization of LuAG SCFs on cheaper YAG and YAP substrates using the BaO-based flux.

Sc-doped garnets

In contrast to Pr^{3+} ions, the Sc^{3+} isoelectronic impurity in YAG and LuAG SCFs has relatively high (0.8-0.55 and ~0.4, respectively) segregation coefficient. This allows to readily achieving the optimum values of Sc doping in SCF of these garnets in the range 0.02-0.045 formula units (f.u) (0.4-0.9 at. %), at which the highest LY of these SCFs is observed (Fig.2b, Table 1).

The Sc^{3+} ions in YAG and LuAG SCF substituted both the dodecahedral {c}- and octahedral (a)- sites of garnet lattice and formed the $Sc_{Y,Lu}$ and Sc_{Al} emission centers, respectively [7]. Such distribution of Sc^{3+} ions over the {c}- and (a)-sites explains the dependence of the positions of emission bands of (Y-Lu)AG:Sc on the Sc content (Fig.2a and 2b, respectively).

(i)   at relatively small (up to 0.4 at. %) Sc content, the band peaked around 265-275 nm prevailed in the emission spectra of YAG:Sc and LuAG:Sc SCFs (Fig.2c). This band is related to the luminescence of excitons localised around Sc ions (LE(Sc) centers).

(ii)   the bands peaked at 313 and 282 nm become dominant in the emission spectra of YAG:Sc and LuAG:Sc SCFs, respectively, at larger (0.4-1.0 at. %) scandium content (Fig.2a and 2c) and that correlates with reaching maximum LY of these SCFs (Fig.2b). These bands can be corresponds to the luminescence of Sc_Y and Sc_{Lu} centers, respectively.

(iii)   the bands peaked at 353 and 310 nm become dominant in the emission spectra of YAG:Sc and LuAG:Sc SCFs, respectively, at the higher then 1.0 at. % Sc content (Fig.2b) where somewhat decreased LY is observed. Theses bands can be related to the luminescence of Sc_{Al} centers.

(iv)   another Sc^{3+}-related bands peaked above 400 and 350 nm occurs at very high (above 3-4 at. %) Sc content in YAG:Sc and LuAG:Sc SCFs and strong decrease of LY is observed in such samples (Fig.1) due to the concentration quenching of the emission of Sc^{3+}-related centers. Most probably these bands corresponds to the luminescence of dimer Sc_{Al}-$Sc_{Y,Lu}$ centers.

The luminescence decay kinetics of the $Sc_{Y,Lu}$ and Sc_{Al} centers is shown in Fig.2d on the example of LuAG:Sc SCF. Under excitation with energy of 6.88 eV in the exciton range the main components of the Sc_{Lu} and Sc_{Al} centers emission show the decay time of 245 and 415 ns (curve 1 and 2, respectively).

The LY of YAG:Sc and LuAG:Sc SCFs depends on the total Sc content and distribution of Sc^{3+} ions over the {c}- and (a)–sites of garnet host. The ratio of the $Sc_{Y,Lu}$ / Sc_{Al} concentrations in these YAG:Sc and LuAG:Sc SCFs mainly depend on the dimension of dodecahedral sites of

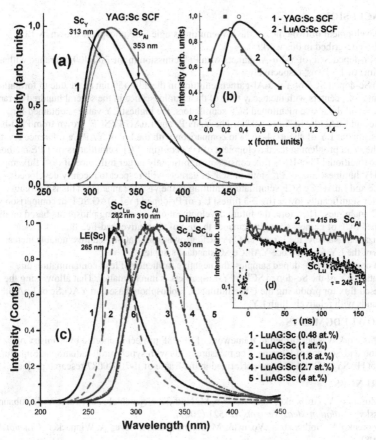

Figure 2. Normalize CL spectra of YAG:Sc (a) and LuAG:Sc (c) SCF with Sc content of 1.8 (1a), 4.13 (2a), 0.48 (1c), 1.0 (2c), 1.8 (3c), 2.7 (4c) and 4 (5c) at.%, respectively. Decomposition of spectrum 2 in Fig.2a is given by dashed lines. (b) - dependence of integral intensity of CL on total scandium content in YAG:Sc (1) and LuAG:Sc (2) SCFs; (d) - decay kinetics of luminescence of Sc_{Lu} (1) and Sc_{Al} (2) centers at 280 nm (1) and 335 nm (2) in LuAG:Sc SCF under excitation by synchrotron radiation with an energy of 6.88 eV. T=300 K.

garnet host. Specifically, this ratio is widely large in LuAG:Sc SCF (~0.5) then in YAG:Sc SCF (~0.3) at optimal (~1-1.8 at. %) scandium content. The LY of (Y-Lu)AG:Sc SCFs at the optimal Sc^{3+} content is comparable with their SC analogues and reaches values of 46-48 % of that for YAP:Ce SC (Table 1). This result shows that influence of Pb^{2+} dopant on the UV luminescence of Sc^{3+}-based centers is not so much significant than that on the Pr^{3+} luminescence in SCF of garnets. This allows the usage of PbO-based flux for the growth of UV- emitting SCF phosphors based on YAG:Sc or LuAG:SC compounds with relatively high LY.

4. CONCLUSIONS

Development of two types of the UV-emitting single SCF phosphors grown by LPE methods is described in this work:

i) Pr-doped SCF of Y-Lu-Al garnets with Pr^{3+} emission in the 300-400 nm range with a decay time of 13-19 ns, respectively;

ii) Sc-doped SCF of Y-Lu-Al-garnets emitting in the 230-450 nm range due to formation of $Sc_{Y,Lu}$ and Sc_{Al} centers with the decay time of their luminescence in the several hundred ns range.

From all the above mentioned SCF scintillators, the highest LY value is obtained for YAG:Sc and LuAG:Sc SCF (46-48 %). The YAG:Pr and LuAG:Pr SCF, grown from PbO-based MS, has smaller LY of about 34-36 % in comparison with the LY of YAP:Ce reference crystal.

The main problem in the development of UV–emitting SCF scintillators by LPE method from the traditional $PbO-B_2O_3$ flux consists in significantly larger influence of Pb^{2+} flux impurity on the UV luminescence of Pr^{3+} ions in SCF of garnets with respect to recently developed YAG:Ce and LuAG:Ce SCF scintillators emitting in the visible range [1]. This is the main reason for significantly lower (by 2-3 times) LY of Pr-doped (Y-Lu)AG SCF in comparison with their SC analogues. Therefore, the future development of UV emitting phosphors based on the Pr^{3+} doped SCF of garnets strongly demands the use of alternative lead-free fluxes for their crystallization. Namely, the LY of YAG:Pr SCF, grown from BaO-based flux, notable increase up to 45 % of the LY for reference YAP:Ce SC standard sample.

In opposite to Pr^{3+} doped samples, the negative influence of lead contamination on scintillation LY of the Sc-doped SCF of Y-Lu-garnets is much smaller. That allow using the PbO-based flux for producing the UV-emitting SCF phosphors based on YAG:Sc or LuAG:Sc compounds with relatively high LY.

ACKNOWLEDGMENTS

This work was performed in framework of SF-28F project supported by Ministry of Education and Science of Ukraine. The measurements with synchrotron radiation at Superlumi station at HASYLAB, DESY was performed in the frame of II-20090087 research project.

REFERENCES

1. Y. Zorenko, V. Gorbenko, E. Mihokova, M. Nikl, K. Nejezchleb, A. Vedda, V. Kolobanov, D. Spassky, *Radiation Measurements* **42**, 521 (2007).
2. Y. Zorenko, V. Gorbenko, T. Voznyak, M. Batentschuk, A. Osvet, A. Winnacker, *J. Lumin.* **128**, 652 (2008).
3. T. Martin, A. Koch, *J. Synchr. Rad.* **13**, 180 (2006).
4. J. Pejchal, M. Nikl, E. Mihóková, J. A. Mareš, A. Yoshikawa, H. Ogino, K. M. Schillemat, A. Krasnikov, A. Vedda, K. Nejezchleb, V. Múčka, *J. Phys. D: Appl. Phys.* **42**, 055117 (2009).
5. N. Ryskin, P. Dorenbos, C. van Eijk, S. Batygov, *J. Phys.: Condens. Matter* **6**, 10423 (1994).
6. V. Gorbenko, A. Krasnikov, M. Nikl, S. Zazubovich, Yu. Zorenko, *Optical materials* **31**, 1805 (2009).
7. Yu. Zorenko, Optics and Spectroscopy **100**, 572 (2006).
8. V. Babin, V. Gorbenko, A. Makhov, J.A. Mares, M. Nikl, S. Zazubovich, Yu. Zorenko, *J. Lumin.* **127**, 384 (2007).
9. Yu. Zorenko, V. Gorbenko, I. Konstankevych, *Surface* **5**, 83 (2003).
10. V. Gorbenko, Yu. Zorenko, V. Savchyn, T. Zorenko, A. Pedan, V. Shkliarskyi, *Optical materials* **45**, 461 (2010).
11. Yu. Zorenko, J.A. Mares, P. Prusa, M. Nikl, V. Gorbenko, V. Savchyn, R. Kucerkova, K. Nejezchleb, *Radiation measurements* **45**, 389 (2010).

Mater. Res. Soc. Symp. Proc. Vol. 1341 © 2011 Materials Research Society
DOI: 10.1557/opl.2011.1480

Bismuth-Loaded Polymer Scintillators for Gamma Ray Spectroscopy

Benjmain L. Rupert, Nerine J. Cherepy, Benjamin W. Sturm, Robert D. Sanner, Zurong Dai and Stephen A. Payne
Lawrence Livermore National Laboratory, Livermore, CA 94550, USA

ABSTRACT

We synthesize a series of polyvinylcarbazole (PVK) monoliths containing varying loadings of triphenyl bismuth as a high-Z dopant and varying fluors, either organic or organometallic, in order to study their use as scintillators capable of gamma ray spectroscopy. A trend of increasing bismuth loading resulting in a better resolved photopeak is observed. For PVK parts with no fluor or a standard organic fluor, diphenylanthracene (DPA), increasing bismuth loading results in decreasing light yield while with samples 1 or 3 % by weight of the triplet harvesting organometallic fluor bis(4,6-difluoropyridinato-N,C2)picolinatoiridium (FIrpic) show increasing light yield with increasing bismuth loading. Our best performing PVK/BiPh$_3$/FIrpic scintillator with 40 wt % BiPh$_3$ and 3 wt % FIrpic has an emission maximum of 500 nm, a light yield of ~30,000 photons/MeV, and energy resolution better than 7% FWHM at 662 keV. Replacing the Ir complex with an equal weight of DPA produces a sample with a light yield of ~6,000 photons/MeV, with an emission maximum at 420 nm and energy resolution of 9% at 662 keV. Transmission electron microscopy studies show that the BiPh$_3$ forms small clusters of approximately 5 nm diameter.

INTRODUCTION

Due to their low cost and ease of fabrication in large volumes plastic scintillators find many uses in radiation detection, such as portal scanners at shipping docks aimed at detecting movements of enriched uranium, plutonium and other potentially dangerous radioactive materials. Scintillator-based gamma ray spectroscopy requires scintillator materials with high light yield, excellent light yield proportionality, prompt emission decay and high effective atomic number for high photoelectric cross-section. Unfortunately plastic scintillators have low light yields, and more importantly, their low effective Z ($Z_{eff} \approx 4.5$) results in poor photopeak efficiency, and they therefore have not been used for gamma-ray spectroscopy. This lack of spectroscopic ability leads to many innocuous shipments of materials, such as ceramics, to be selected for secondary screening, which is time consuming and expensive. Attempts to increase Z_{eff} by doping with heavy metals to induce a photopeak have been made in the past.[1-3] However, difficulties due to poor solubility of the high-Z component and quenching of luminescence were never resolved.

While the low light yields of polymer scintillators impacts their use for event counting to some extent, it becomes critical for gamma ray spectroscopy. Separate from any consideration of spectroscopy, Campbell and Crone produced thin films of PVK doped with an iridium complex fluor which were found to offer improved scintillation light yields.[4] This is due to the same mechanism now widely employed in organic light emitting diodes (OLEDs), wherein it has been found that the Ir-complexes are capable of capturing and emitting both singlet and triplet excitons, due to spin-orbit coupling. While Campbell and Crone demonstrated that higher light yields in thin films are possible with Ir-complex fluors, large scintillator volumes are needed for practical counting rates. Furthermore, if a high-Z component can be introduced, without decreasing the light yield, the capabilities of such a material could go beyond the usual uses for plastic scintillator for gross counting, and be used as a gamma ray spectroscopy scintillator for radioisotope identification applications.

Here we describe our recent efforts to introduce a high-Z component into plastic scintillators with high light yield and good spectroscopic resolution. Triphenylbismuth, long used in the electronics industry as a precursor for inorganic bismuth compounds and therefore readily available,[5] is used as the high-Z material. Rather than thin films of PVK, a material well known in OLED literature as a host material in the active layer,[6-7] we produce monoliths by bulk radical polymerization and incorporate either the high quantum yield organic fluor DPA or the spin-orbit coupling fluor FIrpic.[8-9]

EXPERIMENT

All materials were purchased from Aldrich except FIrpic, which was purchased from American Dye Source and all were used as received, except NVK, which was purified by vacuum sublimation. In a glass ampoule were combined 10 mg Luperox 231, fluorescent or phosphorescent flour as indicated in Table 1, $BiPh_3$ as indicated in Table 1 and enough NVK to bring the total mass to 1.0 g. The ampoule was then placed under vacuum and refilled with nitrogen three times to remove oxygen from the system. While under a slight positive pressure of nitrogen the ampoule was placed in well in a fitted well in an aluminum heating block at 80° C for 24 hours. After cooling again to room temperature the ampoules are broken and the scintillator parts shaped and polished to give right cylinders 18 mm in diameter and 2 mm in height.

Transmission Electron Microscopy (TEM) characterization was demonstrated under a Philips CM300-FEG instrument, operating at 300 kV. The instrument is a fully digital controlled system and has powerful and versatile capabilities including X-ray energy dispersive spectroscopy (EDS) and electron energy loss spectroscopy (EELS) for composition analysis. The TEM specimens were prepared by mechanically grinding small piece of material, and then amounting that on TEM Cu-grids.

Radioluminescence spectra were acquired using a $^{90}Sr/^{90}Y$ source (average beta energy ~1 MeV), and were collected with a Princeton Instruments/Acton Spec 10 spectrograph coupled to a thermoelectrically cooled CCD camera and corrected for spectral sensitivity. Gamma ray spectra were acquired using a Hamamatsu R6231-100 superbialkali PMT. Plastic scintillators were centered on the entrance window, optically coupled to the PMT with optical grease and wrapped with several layers of Teflon tape. The PMT signals were shaped with a Tennelec TC 244 spectroscopy amplifier (8 μs shaping time for FIrpic activated samples and 1 μs for DPA activated samples) then recorded with an Amptek MCA8000-A multi-channel analyzer.

DISCUSSION

As described earlier, attempts had been made in the past to incorporate high-Z dopants, for example BiPh₃, into polymer scintillators in order to obtain a photopeak and enable gamma ray spectroscopy.[1-3] The failure of these materials to obtain satisfactory performance were never fully explained in the literature, but can be understood if the relative excited state energies of the various materials are considered. Keeping with the examples of BiPh₃ and polyvinyl toluene (PVT) the HOMO-LUMO gaps are 4.1 eV[11] and 4.2 eV[11] respectively. Energy transfer between the two is likely facile (fig 1a) and if the organometallic species either: (1) introduces a fast exciton relaxation pathway, or, (2) have a fast intersystem crossing to the triplet exciton state (a non-emissive state for the organic fluors in these systems), the light yield of the resulting scintillators will be significantly diminished.

These limitations in the PVT/BiPh₃ polymer scintillator system can be overcome with proper materials selection. The improved light yields reported by Campbell and Crone in the PVK/Ir-dye scintillators suggest such materials. PVK has a smaller HOMO-LUMO gap (3.5 eV)[10] than BiPh₃ and energy transfer from the organometallic dopant to the matrix is expected to be competitive with relaxation of the excited state and energy back transfer to the Bi dopant should not occur. Use of a high quantum yield phosphorescent dye should also improve the brightness of the scintillator while solving the possible problem of increased intersystem crossing induced by the heavy metal dopant. Care must be taken with the selection of the Ir dye however as the photomultiplier tubes used in gamma ray spectroscopy have a sharp drop off in sensivity starting around 500 nm. FIrpic (fig 1c) was chosen as it was the shortest wavelength Ir dye commercially available when we began this study.

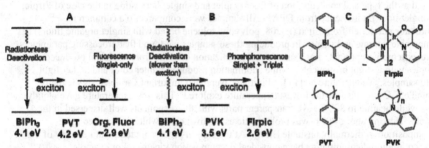

Figure 1. **(A)** Energy levels of Triphenyl Bismuth (BiPh₃), Polyvinyltoluene (PVT) and a typical organic fluor, Diphenyl Anthracene (DPA). **(B)** Energy levels of BiPh₃, Polyvinylcarbazole (PVK) and an Iridium complex fluor (FIrpic). **(C)** Chemical structures of the constituent materials used in the high-Z polymer scintillators described in this work.

Table 1. Composition and summary of spectroscopy results of the samples used in this study.

Sample	Matrix	Fluor	$BiPh_3$ Wt. %	βLY (Ph/MeV)	Relative Gamma Light Yield*	Resolution FWHM*
1a	PVK	None	0%	1420	-¥	-¥
1b	PVK	None	25%	-$	-¥	-¥
1c	PVK	None	40%	-$	-¥	-¥
2a	PVK	3% DPA	0%	10120	-¥	-¥
2b	PVK	3% DPA	25%	6979	-¥	-¥
2c	PVK	3% DPA	40%	6281	0.66	9%
3a	PVK	1% FIrpic	0%	17400	-¥	-¥
3b	PVK	1% FIrpic	25%	22069	-¥	-¥
3c	PVK	1% FIrpic	40%	25464	-¥	-¥
4a	PVK	3% FIrpic	0%	24191	0.73	9%*
4b	PVK	3% FIrpic	25%	25464	-¥	-¥
4c	PVK	3% FIrpic	40%	30641	0.78	6.8 %
Eljen EJ208	PVT	Proprietary	0%	17000	1.0	8%*

At 662 keV   **From fit to Compton edge   ¥Not Measured $Value too small to measure

With these concepts in mind we made a series of twelve scintillator samples (Table 1) containing 40 wt%, 25 wt% or no $BiPh_3$ and either no fluor, 3 wt% DPA or 1 wt% or 3 wt% FIrpic. By examining this series we can separate out the effects of weight percent of Bi dopant as well as the type and concentration of fluor (triplet and singlet harvesting in the case of FIrpic, or singlet-only fluorescence from DPA). All samples were compared to a commercial scintillator purchased from Eljen (EJ208, polyvinyl toluene based with singlet organic fluors) of comparable volume and shape. In producing these samples we found that processing parameters were extremely important. For example recrystallization of NVK from methanol produced samples with a slight brown color, while sublimation produced lighter color and higher light yield samples. Temperature and radical initiator were also important variables. NVK melts around 65° C and this was the lowest temperature explored. This low temperature gave cloudy parts while heating to 80° C gave transparent parts with all components well dispersed in the sample. Heating above 80° C was problematic as a brown color which is assumed to be decomposition of thermally unstable $BiPh_3$. At the chosen processing temperature many of the most common radical initiators have non-ideal decomposition kinetics. For example azobisisobutyronitrile (AIBN) decomposes too quickly at this temperature resulting in a poorly controlled polymerization and bubbles in the solidified matrix. In contrast many common peroxide such as tert-butyl peroxide decompose slowly and polymerization is often not complete for days. Luperox 231 allowed polyermization to be completed in 24 hours while still giving mechanically robust and bubble-free samples.

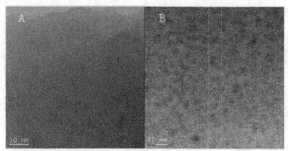

Figure 2. High resolution TEM of **(A)** Sample 3a and **(B)** 3c. White scale bars are 10 nm and red bars are 5 nm.

Before beginning scintillation studies with these samples we examined some by transmission electron microscopy (TEM) as high resolution phase contrast images can give us information about the dispersion state of BiPh₃. TEM phase contrast images are presented in figure 2. Fig 2a, which is an image of sample **3a** (no BiPh₃) shows no obvious phase-separation, where as in sample **3c**, (40 wt% BiPh₃) we find high-Z clusters even dispersed throughout the samples with diameter of about 5 nm (fig 2b). Samples with which contain BiPh₃ at lower loadings (not shown) also showed high-Z clusters of the same diameter. While this is larger than a single BiPh₃ molecule, it likely represents a cluster of only several molecules. Attempts to co-polymerize the Bi-dopant by using a vinyl derivative in either PVK or PVT resulted in similar clustering (not shown).

We began radioluminescence studies by making beta light yield (βLY) measurements on these samples. In this experiment the samples exposed to high energy electrons, similar to the photoelectron ejected from a core orbital of a high-Z element during gamma spectroscopy, and the luminescence spectra are recorded. A spectrum is obtained, rather than counts with a photomultiplier tube (PMT), and is sensitive to the full range of wavelengths emitted by all of the dyes studied. This allows the samples to be checked for luminescence intensity without regard to spectral match to the PMT or to photoelectron efficiency as would be the case in gamma spectroscopy measurements. The first samples measured, **1a-c**, have low light intensity as expected as they have no flour and the quantum fluorescence quantum yield of PVK is low. In both the 25 and 40 wt% BiPh₃ samples light yields were below the detection limit. This indicates that there is still some quenching by the Bi-dopant from either a fast relaxation pathway or intersystem crossing to a triplet state. In samples **2a-c** with DPA and varying BiPh₃ loading, while light yields are substantially improved over the corresponding series 1a-c., the light yields still fall with increasing BiPh₃ concentration.

All samples in series **3a-c** and **4a-c** are brighter in βLY measurements than EJ208 and the brightness vs. BiPh₃ loading trends are different. Each sample in series **4**, with 3 wt% FIrpic, is brighter than the corresponding sample in series **3** with only 1 wt% Ir dye, albeit by a factor of less than 2x, indicating that not all excitons are being collected by the fluor in the lower concentration series. Higher concentrations of FIrpic have not yet been tested for higher beta light yields. Within each series **3** and **4**, and in contrast to series **1** and **2**, higher light yields are observed with increasing BiPh₃ concentrations. Although the mechanism behind this phenomenon is not yet known, all observations so far are consistent with the hypothesis that the higher BiPh₃ concentrations are resulting in a larger percentage of triplets in the system. This would reduce the light yield in the singlet-only emitters but might increase the light yields of triplet emitters if the triplet excitons were either more mobile or longer lived. Time resolved spectroscopy studies will hopefully allow a full mechanism to be elucidated in the future.

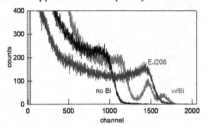

Figure 3. Cs-137 pulse height spectra of EJ208, 4a and 4c.

A subset of the samples were chosen for gamma ray spectroscopy using a Cs-137 source (662 keV gamma energy). Samples **2c**, **4a**, **4c** and EJ208 were compared in order to determine the effect of Bi loading and the use of singlet vs. triplet fluor (**4a**, **4c** and EJ208 shown in fig 3). Besides looking for a photopeak we also determined gamma light yields by direct comparison of the Compton edge with that of the EJ208 commercial sample, which was taken to have a light yield of 1. The DPA sample **2c** (40 wt% $BiPh_3$), which had a significantly lower βLY than than EJ208, has a gamma light yield of 0.66. The samples **4a** and **4c**, with the triplet fluor and 0 and 40 wt% $BiPh_3$, respectively, both had higher βLY than EJ208, but have relative gamma light yields of 0.73 and 0.78. This lower light yield can be attributed to the spectral mismatch between the fluor and the PMT response. The increased light yield with added bismuth dopant in this case agrees with the βLY observation.

The Cs-137 pulse height spectra reveal that the two $BiPh_3$ doped samples, **2c** and **4c**, have a photopeak. Sample **2c** has a resolution of 9% while the higher light yield **4c** has a resolution of 6.8% at 662 keV. Sample **4c**'s resolution is close to that of NaI (6%), which is a commonly used scintillator material. In both of these samples two peaks are present, a full energy peak and an escape peak corresponding to the loss of a 77 keV x-ray generated when an electron falls in to the vacancy in the core Bi shell created by the photo electron. This escape peak is greater in area than the full energy peak due to the moderate Z_{eff} of the samples and their small volume. In samples of sufficient volume, without any change in sample formulation, only the full energy peak should be present. The two samples lacking $BiPh_3$ do not have a photopeak, clearly demonstrating the need for the high-Z dopant for spectroscopy applications.

CONCLUSIONS

In summary, polyvinylcarbazole-based plastic scintillators loaded with up to 40% triphenyl bismuth and activated with standard singlet fluors as well as spin-orbit coupling fluors may be formed with excellent mechanical integrity and high transparency. The light yields obtained with the Iridium complex spin-orbit coupling fluor (FIrpic) is about 3x that of the standard fluor, and the energy resolution with this sample is superior, as well. Energy resolution measured for the FIrpic-activated sample 662 keV is 6.8%, close to that of NaI. These initial results acquired with small samples are very promising, and scaleup to larger volumes is in progress. Large polymer scintillators offering equivalent performance to those reported here would have a wide range of uses, potentially replacing single crystals for some applications, and expanding the capabilities of plastic scintillators in applications where they are currently employed.

ACKNOWLEDGMENTS

This work was supported by the National Nuclear Security Administration, Office of Defense Nuclear Nonproliferation, Office of Nonproliferation Research and Development (NA-22) of the U.S. Department of Energy under Contract No. DE-AC03-76SF00098, and performed under the auspices of the U.S. Department of Energy by Lawrence Livermore National Laboratory under Contract DE-AC52-07NA27344.

REFERENCES

1. S. R. Sandler and K. C. Tsou, Int. J. Appl. Radiat. Is. 15, 419 (1963).
2. K. C. Tsou, IEEE T. Nucl. Sci., 28 (1965).
3. J. Dannin, S. R. Sandler and B. Baum, Int. J. Appl. Radiat. Is. 16, 589 (1965).
4. I. H. Campbell and B. K. Crone, Appl. Phys. Lett. 90, 012117 (2007).
5. M. Mehring, Coord. Chem. Rev., 251, 974 (2007).
6. N. Tamaki, Polymer J., 42, 103 (2010)
7. J. V. Grazulevicius, P. Strohriegl, J. Pielichowski and K. Pielichowski, Progr. Polym. Sci., 28, 1297 (2003).
8. S. Lamansky, P. Djurovich, D. Murphy, F. Abdel-Razzaq, R. Kwong, I. Tsyba, M. Bortz, B. Mui, R. Bau and M. E. Thompson, Inorg. Chem., 40, 1704 (2001).
9. R. J. Holmes, S. R. Forrest, Y.-J. Tung, R. C. Kwong, J. J. Brown, S. Garon and M. E. Thompson, Appl. Phys. Lett., 82, 2422 (2003).
10. F. J. Zhang, Z. Xu, D. W. Zhao, S. L. Zhao, L. W. Wang and G. C. Yuan, Phys. Scr. 77, 055403 (2008).
11. Estimated from UV-vis.

Mater. Res. Soc. Symp. Proc. Vol. 1341 © 2011 Materials Research Society
DOI: 10.1557/opl.2011.1482

Alkali Metal Chalcogenides for Radiation Detection

J. A. Peters[1], Zhifu Liu[1], and B. W. Wessels[a), 1,2]
I. Androulakis[3], C. P. Sebastian[3], Hao Li[3], and M. G. Kanatzidis[3]

1 Materials Research Center, Northwestern University, Evanston, IL 60208

2 Department of Electrical Engineering and Computer Science, Northwestern University, Evanston, IL 60208

3 Department of Chemistry, Northwestern University, Evanston, IL 60208

ABSTRACT

We report on the optical and charge transport properties of novel alkali metal chalcogenides, $Cs_2Hg_6S_7$ and $Cs_2Cd_3Te_4$, pertaining to their use in radiation detection. Optical absorption, photoconductivity, and gamma ray response measurements for undoped crystals were measured. The band gap energies of the $Cs_2Hg_6S_7$ and $Cs_2Cd_3Te_4$ compounds are 1.63 eV and 2.45 eV, respectively. The mobility-lifetime products for charge carriers are of the order of $\sim 10^{-3}$ cm^2/V for electrons and $\sim 10^{-4}$ cm^2/V for holes. Detectors fabricated from the ternary compound $Cs_2Hg_6S_7$ shows well-resolved spectroscopic features at room temperature in response to γ-rays at 122 keV from a ^{57}Co source, indicating its potential as a radiation detector.

INTRODUCTION

The need for nuclear radiation detectors with good spectroscopic performance that operate at room temperature has led to a widening search for new wide gap semiconductors. To date, several semiconductors have had significant impact on radiation detection that include Ge, HgI$_2$, CdTe, and Cd$_{1-x}$Zn$_x$Te[1,2]. However, there are a number of limitations in the use of these semiconductors. For example, Ge requires cryogenic cooling to function as a radiation detector. It also lacks sufficient sensitivity to gamma rays despite its impressive performance because of its low density and low atomic number[3]. Furthermore, while Cd$_{1-x}$Zn$_x$Te is currently a leading material for room temperature radiation detectors, growth of large homogeneous crystals remains a challenge.

Thus it is evident that new materials are needed for advanced radiation detectors that operate at room temperature. The choice of suitable semiconductor materials for radiation detection is mainly determined by the energy range of interest. The selection criteria for these materials relevant for x-ray and γ-ray detection include high atomic number, high density, and wide energy band gaps in the range $1.6 < E_g < 2.6$ eV. Moreover, these materials must work at room temperature. However, compound semiconductors with a high atomic number typically have band gaps of less than 1 eV. It has been previously shown that dimensional reduction can be exploited to generate new ternary compounds with desired energy band gaps. These compounds include A$_2$Hg$_6$Q$_7$ (A=Cs, Rb, K; Q=S, Se), A$_2$Hg$_3$Q$_4$ (A=Cs, K; Q=S, Se), and A$_2$Cd$_3$Q$_4$ (A=Cs, K; Q=S, Se, Te)[4,5,6].

Here, we present the optical and charge transport properties of two alkali heavy metal chalcogenides $Cs_2Hg_6S_7$ and $Cs_2Cd_3Te_4$ grown using a modified Bridgman method. These chalcogenides are promising candidate materials for radiation detection because of their high atomic number elements (Z_{Cs} = 55, Z_{Hg} =80, Z_S = 16, Z_{Cd} = 48, Z_{Te} = 52) and relatively high densities ($d_{Cs_2Hg_6S_7} = 6.94\,g/cm^3$ and $d_{Cs_2Cd_3Te_4} = 5.54\,g/cm^3$, respectively). In particular, optical absorption measurement is used for the derivation of the energy band gaps of these materials.

Band gap energies of 1.61 eV and 2.45 eV were measured for $Cs_2Hg_6S_7$ and $Cs_2Cd_3Te_4$, respectively. Steady state photoconductivity was used to characterize and measure the charge transport properties. Detectors were fabricated and their spectroscopic performance was assessed at room temperature.

EXPERIMENT

The ternary compounds $Cs_2Hg_6S_7$ and $Cs_2Cd_3Te_4$ were grown using the modified Bridgman technique. Details of the growth and synthesis of these materials are reported elsewhere.[5,7] Phase identity and purity were determined by powder X-ray diffraction experiments using Cu Kα radiation. Energy-dispersive X-ray spectroscopy (EDS) analysis taken on visibly clean surfaces of the samples gave the atomic composition of $Cs_2Hg_6S_7$ and $Cs_2Cd_3Te_4$. The samples were first mechanically polished with abrasive paper and then with 0.05 alumina suspension. The crystals were then rinsed in ethanol. The samples had thicknesses of ~1-3 mm. The resistivity of the crystals was ~10^6 Ω-cm as measured by current-voltage (I-V) characteristics.

The optical transmittance $T(\lambda)$ and reflectance $R(\lambda)$ of the crystals were measured at normal incidence in the wavelength range of 300 – 1500 nm using a Lambda 1050 UV-Vis-near IR spectrophotometer. The dependence of absorption coefficient,α on wavelength was determined for the case of allowed direct (α^2) optical transitions.

Photoconductivity measurements at room temperature were carried out on parallel plate samples that were illuminated with photons with 633 nm wavelength. The mobility-lifetime ($\mu\tau$) products for electrons and holes were obtained by applying either negative or positive voltage to the illuminated detector, and measuring the photocurrent for electrons and holes. For radiation testing, we fabricated ~ 3 mm thick slab of $Cs_2Hg_6S_7$ and ~ 1 mm thick slab of $Cs_2Cd_3Te_4$ into detectors. Thin Ti/Au electrodes were deposited on opposite faces in a parallel plate configuration. The detectors were then exposed to ^{57}Co γ-rays and the spectra measured using photon counting instrumentation.

RESULTS and DISCUSSION

Optical properties and determination of the optical band gap
The absorption coefficient of the compounds was determined by transmission measurements. By averaging over multiple reflection effects, the values of transmission T are determined using the following equations for bulk semiconductors[8]

$$T = (1-R)^2 \exp(-\alpha d), \tag{1}$$

$$\alpha = \log\left[\left((1-R)^2/2T\right) + (1-R)^4 / \left(4T^2 + R^2\right)\right]^{\frac{1}{2}} \tag{2}$$

where R and α are the reflectivity and the optical absorption coefficient respectively, and d is the thickness of the sample.

The spectral dependence (in the wavelength range of 300-1500 nm) on T and R for $Cs_2Hg_6S_7$ and $Cs_2Cd_3Te_4$ is given in Fig. 1. We observe in fig. 1(a) that $Cs_2Hg_6S_7$ becomes transparent in the infrared region, whereas $Cs_2Cd_3Te_4$ becomes transparent for $\lambda > 500$ nm.

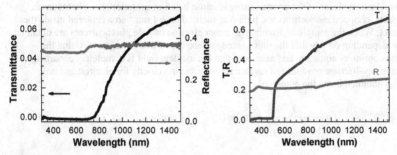

Figure 1: (a) Spectral distribution of transmittance and reflectance for $Cs_2Hg_6S_7$, in the wavelength range of 300-1500 nm. **(b)** Spectral distribution of transmittance and reflectance for $Cs_2Cd_3Te_4$, in the same wavelength range of 300-1520 nm.

The dependence of absorption coefficient,α on photon energy was measured and analyzed for the case of allowed direct optical transitions as shown in Fig. 2 for the two compounds. The energy band gaps were determined by extrapolating the linear portion of the α^2 curve towards lower photon energies and determining the point of interception with the energy axis. The values for the direct energy gap for $Cs_2Hg_6S_7$ and $Cs_2Cd_3Te_4$ are 1.63 eV and 2.45 eV, respectively. These values are consistent with those previously measured for the series of $A_2Hg_6Q_7$ and A_2Cd_3Q4 compounds, which range from 1.6 - 1.7 eV and 2.26 eV - 2.92 eV.[4,5] The bandgap energies are indeed higher than the bandgap energy of 1.2 eV for β-HgS and 1.5 eV for CdTe. These findings support the dimensional reduction paradigm.[4, 5, 6]

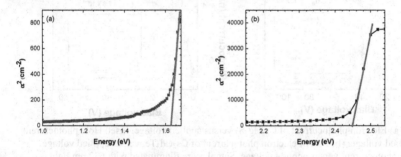

Figure 2: Dependence of (α^2) on photon energy (hv) for **(a)** $Cs_2Hg_6S_7$. **(b):** $Cs_2Cd_3Te_4$.

Steady-state photoconductivity

To determine the charge transport properties, photoconductivity of $Cs_2Hg_6S_7$, and $Cs_2Cd_3Te_4$ was measured. The thickness of the $Cs_2Hg_6S_7$ sample was 2 mm while that of $Cs_2Cd_3Te_4$ was ~ 0.85 mm. Thin film, transparent gold electrodes of ~80 nm were deposited onto the front and back surfaces of the crystals using electron beam evaporation. For photocurrent measurements, chopped monochromatic light was focused on a 4 mm^2 area centered inside the metal contact. When the sample is illuminated, either electron or hole photocurrents are obtained separately, depending on whether the illuminated surface is positive or negative. Using the steady-state continuity equation, and assuming that the incident light is completely absorbed, both the mobility-lifetime product and surface recombination velocity for electron and hole carriers was obtained using the equation[9]

$$I = \frac{I_0}{1+\dfrac{L}{V}(s/\mu)_{e,h}}(\mu\tau_b)_{e,h}(V/L^2)\left[1-\exp\left(-L^2/V(\mu\tau_b)_{e,h}\right)\right] \tag{3}$$

Here, L is the detector thickness, V is the applied bias, s is the surface recombination velocity, μ is the carrier mobility, and τ_b is the bulk lifetime. I_0 is the saturation photocurrent at high bias is proportional to the carrier generation rate. Equation 3 has been successfully employed to analyze charge transport including the surface generation and recombination effects in the photoconductivity of other detector materials.[10] Figure 3(a) shows typical curves of the electron and hole photocurrent versus applied voltage measured for $Cs_2Hg_6S_7$ at room temperature. The solid lines are theoretical fits to Eqn. 3. Here $\mu\tau_b$ and s/μ are fitting parameters. The $\mu\tau_b$ values are 1.92×10^{-3} cm^2/V and 3.42×10^{-4} cm^2/V for electrons and holes, respectively. The s/μ values are 44 and 118 V/cm, respectively for the front and back surfaces of the sample.

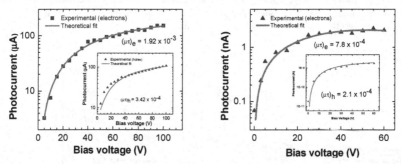

Figure 3: (a) Electron photocurrent of $Cs_2Hg_6S_7$ versus applied voltage. Inset: Hole photocurrent versus applied voltage. **(b)** Plot of electron photocurrent of $Cs_2Cd_3Te_4$ versus applied voltage. Inset: Hole photocurrent versus applied voltage. Samples are illuminated with 633 nm light.

Figure 3(b) shows curves of electron and hole photocurrents versus applied voltage for $Cs_2Cd_3Te_4$ at room temperature. Since the measured photocurrent was obtained with sub-band-gap illumination, the incident light is weakly and uniformly absorbed through the crystal. Thus the mobility-lifetime product and surface recombination velocity were obtained by fitting the experimental result with the equation[10] (the fit is shown by the solid lines in Fig. 3b):

$$I = 2I_0\left(V/L^2\right)\left(\mu\tau_b\right)_{e,h}\left\{1-\left(V/L^2\right)\left(\mu\tau_b\right)_{e,h}\left[1-\exp\left(-L^2/V\left(\mu\tau_b\right)_{e,h}\right)\right]\right\} \tag{4}$$

When a detector is illuminated with below-band-gap light, photo-generated carriers are produced by ionization of defects or traps. The measured photocurrent therefore pertains to the bulk defect/trap ionization mechanism and transport of carriers through the bulk.[11] Nevertheless, it has been shown that charge collection efficiency values measured from both photoconductivity for below-band-gap light and high-energy γ-ray detector response are equivalent.[12] The values of $(\mu\tau_b)_e$ and $(\mu\tau_b)_h$ obtained by fitting the experimental data in fig.3b to Eqn. 4 are 7.8×10^{-4} cm^2/V and 2.1×10^{-4} cm^2/V, respectively. It is worth noting that the measured optical, electrical, and detector parameters of these alkali metal chalcogenides compare favorably with that of other room temperature detector materials.[13,14] Specifically, the mobility-lifetime products measured from steady-state photoconductivity are higher than those reported for TlBr, PbI$_2$ and HgI$_2$[15] although the $(\mu\tau)_{e,h}$ values for Cs$_2$Cd$_3$Te$_4$ are less than those of CdTe.

Detector performance

Parallel plate Cs$_2$Hg$_6$S$_7$, and Cs$_2$Cd$_3$Te$_4$ detectors were exposed to radiation from a ^{57}Co source and the energy spectrum data was measured using a multichannel analyzer with MAESTRO software. The resistivity of the undoped crystals were 5×10^6 Ω-cm and 1×10^6 Ω-cm, respectively. The Cs$_2$Cd$_3$Te$_4$ crystal did not exhibit any response to the radiation due to its low resistivity and sample thickness (~ 0.85 mm). The Cs$_2$Hg$_6$S$_7$ crystal however was responsive to the ^{57}Co radiation. The peak in the spectrum [fig. 4(a)] could be distinguished over the noise and background although the peak is not very well resolved. For comparison, the ^{57}Co spectrum data was taken [fig. 4(b)] for CdZnTe (10% Zn) detector. We see from fig. 4(a) that only the 122 keV peak is resolved for Cs$_2$Hg$_6$S$_7$. This is presumably due to the low resistivity of Cs$_2$Hg$_6$S$_7$, which results in a large dark current and a broad energy peak.

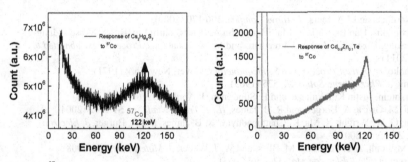

Figure 4: ^{57}Co spectrum measured with **(a)** Cs$_2$Hg$_6$S$_7$ and **(b)** Cd$_{0.9}$Zn$_{0.1}$Te detectors.

SUMMARY

We have measured the optical and transport properties of suitable photoconductive detectors based on $Cs_2Hg_6S_7$, and $Cs_2Cd_3Te_4$ and the results have been compared to other semiconductor materials that have had significant impact on radiation detection. Direct energy gaps of 1.63 eV and 2.45 eV were determined for undoped $Cs_2Hg_6S_7$, and $Cs_2Cd_3Te_4$, respectively. Using the photoconductivity measurements the materials exhibited good transport properties with $(\mu\tau)_e$ ~10^{-3} and $(\mu\tau)_h$ ~ 10^{-4}. These compounds exhibit high mobility lifetime products directly comparable to optimized TlBr, HgI_2, and $Cd_{0.9}Zn_{0.1}Te$. The $Cs_2Hg_6S_7$ detector shows well-resolved spectroscopic features in response to a ^{57}Co γ-ray source at 122 keV, indicating its potential as a promising candidate for applications in radiation detection.

ACKNOWLEDGMENTS

This work was supported by DTRA under grant no. HDTRA1-09-1-0044. Extensive use of the microfabrication facilities of the Materials Research Center at Northwestern University supported by the NSF (No. DMR 0076097) is acknowledged.

REFERENCES

[1] A. Owens, and A. Peacock, "Compound semiconductor radiation detectors", Nucl. Instr. & Meth. in Phys. Res., Vol. **531**, pp 18-37 (2004)

[2] T. E. Schlesinger, J. E. Toney, H. Yoon, E. Y. Lee, B. A. Brunett, L. Franks, and R. B. James, "Cadmium zinc telluride and its use as a nuclear radiation detector material," *Mater. Sci. Eng.*, Vol. **32**, p. 103, 2001.

[3] P. N. Luke, and M. Amman, *15th International Workshop on Room-Temperature Semiconductor X- and Gamma-Ray Detectors*, San Diego, Ca, 2006.

[4] E. A. Axtell, J. H. Liao, Z. Pikramenou, and M. G. Kanatzidis, *Chem. Eur. J.* **2** 656 (1996)

[5] E. A. Axtell, Y. Park, K. Chondroudis, and M. G. Kanatzidis, *J. Am. Chem. Soc.* **120** 124 (1998)

[6] A.A. Narducci and J.A. Ibers, *J. Alloys Compds.* **306** 170 (2000).

[7] J. Androulakis, Hao Li, Christos Malliakas, John A. Peters, Zhifu Liu, Bruce Wessels, Jung-Hwan Song, Hosub Jin, Arthur J. Freeman, and M. G. Kanatzidis, *Proceedings submitted to MRS* (2011)

[8] J. I. Pankove, *Optical Processes in Semiconductors*, Dover, New York (1971).

[9] A. Many, *J. Phys. Chem. Solids* **26**, 575 (1965).

[10] Z. Burshtein, Justin K. Akujieze, and E. Silberman, J. Appl. Phys. **60**, 3182 (1986)

[11] Y. Cui, M. Groza, A. Burger, and R. B. James, *IEEE Trans. Nucl. Sci.* **51** 1172 (2004).

[12] Y. Cui, G. W. Wright, X. Ma, K. Chattopadhyay, R. B. James, and A. Burger, *J. Electron. Mater.* **35** 1267 (2003).

[13] B. D. Milbrath, A. J. Peurrung, M. Bliss, and W. J. Weber, *J. Mater. Res.*, **23**, (2008).

[14] A. Burger et. al., *J. Electron Mat.* **32**, 756 (2003).

[15] Del Sordo S. et al., Sensors, **9** 3491 (2009).

Mater. Res. Soc. Symp. Proc. Vol. 1341 © 2011 Materials Research Society
DOI: 10.1557/opl.2011.1110

ZnCdSeTe Semiconductor Compounds: Preparation and Properties

Vello Valdna[1], Maarja Grossberg[1], Jaan Hiie[1], Urve Kallavus[2], Valdek Mikli[2], Taavi Raadik[1], Rainer Traksmaa[2] and Mart Viljus[2]
[1]Department of Materials Science, Tallinn University of Technology, 5 Ehitajate Rd., 19086 Tallinn, Estonia
[2]Centre for Materials Research, Tallinn University of Technology, 5 Ehitajate Rd., 19086 Tallinn, Estonia

ABSTRACT

Group II-VI narrow band gap compounds CdTe, ZnCdTe and CdSeTe are known as the most suitable semiconductor materials for the room temperature gamma- and X-ray radiation detectors. In this work, we investigated electronic properties of a quaternary compound ZnCdSeTe. Cl, Cu, Pr, Er and oxygen doped host materials were synthesized from the grinded mixture of 6N purity ZnTe, CdTe and CdSe by the help of $CdCl2$ flux. Precautions were applied to achieve an uniform doping and high quality of the crystal surfaces. Residue phases after the thermal treatments were removed by the help of a vacuum annealing. It was found that Zn increases a substitutional solubility of dopants in ZnCdSeTe and thus, promotes optoelectronic properties of the ZnCdSeTe alloy. Cl substitutes Te whereas Cu and rare earth elements substitute Zn in ZnCdSeTe. Fabricated polycrystalline samples showed a high performance from NIR via VIS and UV to X-ray band. High stability, good linearity and performance of samples was measured under X-ray excitation of Cu Kα 1.54056 Å, at 40 kV.

INTRODUCTION

Group II-VI narrow band gap compounds like CdTe, ZnTe, CdSe and its solid solutions are widely used electronic materials. CdTe is the most frequently used semiconductor material used for room temperature γ- and X-ray detectors [1]. Chlorine- doped CdTe is high- or low-resistivity p-type material, depending on the chlorine concentration [2]. Clorine- and oxygen-doped [3] or chlorine- copper- and oxygen-doped [4] CdTe has a high resistivity, and a negligible photoconductivity. Alloys of CdTe-CdSe containing 30 to 45 mol% CdSe could be made either cubic or hexagonal structure depending on the annealing temperature and cooling rate [5]. Chlorine-, copper- and oxygen-doped CdSe- and CdSeTe- based wurtzite structures have a high photoconductivity whereas a photoconductivity of zincblende CdTe and CdSeTe is much lower.

Alloying CdTe with Zn increases the resistivity from 1E9 to 1E10-1E11 Ωcm [6, 7]. CdZnTe has emerged as one of the most attractive and promising materials for room-temperature γ- and X-ray spectroscopy, for medical imaging, national security, environmental monitoring, and space astronomy [8]. Detectors based on Si or Ge material can only work efficiently at liquid-nitrogen temperature, at high X-ray photon energies, the use of Si detectors is greatly limited because of the reduced stopping power [9]. The energy required for generating one electron-hole pair in CdZnTe (~5 eV) is much less than that required for scintillation crystals coupled to photomultiplier tubes (~50 eV), resulting in better energy resolution [8]. The primary advantages of the resistive detectors for good quality imaging are: efficient radiation absorption, large band gap energy limiting the thermal generation of charge carriers in the bulk, good linearity, good charge transport properties, high stability, high sensitivity and wide dynamic range [7]. Despite the continuous efforts made in the material development, uniform and large-area CdTe and CdZnTe bulk crystal growth is still very difficult. Hence, development of large area imaging arrays has been a major challenge [10].

The aim of this paper is to investigate optoelectronic properties of the doped quaternary compounds ZnCdSeTe and Zn influence on the properties of these alloys. International Centre for Diffraction Data (ICDD) gives mass attenuation coefficients for Cu Kα of elements Zn, Cd, Se and Te 60.3, 231, 91.4 and 282 cm^2/g, respectively. Thus, these compounds are well suitable for high-energy radiation detectors.

EXPERIMENT

Host materials were synthesized from the 6N purity grinded mixture of ZnTe, CdTe, and CdSe. Se/Te ratio 31/19 at% was used for the all synthesized materials to keep a hexagonal structure of crystals [5] whereas Zn/Cd ratio was changed from 0/50 at% to 10/40 at%. 5N purity CdCl$_2$ and CuCl$_2$ were used as a flux and a source of Cl and Cu dopants, and 5N purity PrCl$_3$ and ErCl$_3$ as a source of Pr and Er co-dopants at part of materials. Cu concentrations in the range of 0.008-0.05 at% and Pr and Er concentrations up to 0,035 at% were used. Synthesis was carried out in a Lindberg/Blue M three-zone tube furnace at a temperature of 900 ^{0}C, inside the vacuum sealed quartz ampoule. Continuous rotation during the synthesis in the slightly tilted quartz reactor was used to better homogeneity. From the synthesized material were pressed 4 mm pellets, with a thickness of 0.6 mm. Pellets were recrystallized inside the vacuum sealed quartz ampoules at 550 ^{0}C for one hour. Formed at the thermal treatments flux residue and excess phases were removed by the vacuum annealing of samples inside a long two-zone ampoule at temperatures up to 570 ^{0}C. One end of the ampoule was kept at 300 K, and connected to the vacuum system with a residual pressure below 1E-2 Torr. Annealing duration from one hour to some days was used. Oxygen doping was carried through by the annealing of samples in an oxygen ambient at temperatures up to 550 ^{0}C using a method described in [3]. For conductivity measurements indium electrodes were evaporated onto the samples.

Photoluminescence (PL) spectra of samples were measured in the Janis CC-150 closed-cycle He cryostat. The 441.6 nm line of Omnichrome He-Cd laser was used as the excitation source of samples. A chopped signal from the Hamamatsu InGaAs photodetector was amplified, recorded and corrected on a computer.

Condensed phases after the thermal treatment, crystal structure and surface morphology of the samples were studied using Bruker AXS, Inc. X-ray diffractometer D5005 and Zeiss/LEO Supra 35 apparatus equipped with the Thermo Maxray Extended Range WDX detector with accuracy of 0.02 mass%. Chemical composition of the condensed phases was evaluated by energy-dispersive X-ray spectroscopy (EDS) analysis in the Link Analytical AN10000 analyser. Dark resistance R$_D$ and saturated light resistance R$_L$ near a glow lamp of 60 W were measured from the 2x2 mm surface of the pellets using teraommeter E6-13.

DISCUSSION

Figure 1 presents PL spectra of two ZnCdSeTe recrystallize d samples having a different Zn concentration but the same dopants concentration: [Cu] = 0.04 at%, [Pr] and [Er] = 0.035 at%, and [Cl] – saturated. Sample 1 has a higher zinc concentration [Zn] = 10 at% and higher PL intensity whereas at sample 2 [Zn] = 0.5 at%. Dark resistance of recrystallized samples R_D is around 1E5-1E7 Ω, photoconductivity is poor, with dark to light resistance ratio K_F = 1-10. EDS detected Te, Zn and $CdCl_2$ precipitates on the surface of recrystallized samples. Vacuum annealed 1 hr at 340 ^{0}C samples have R_D = 1E6-1E8 Ω, and K_F = 1-2.

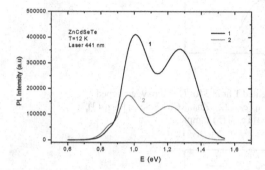

Figure 1. PL spectra of the recrystallized 1 hr at 550 ^{0}C $Zn_{0.1}Cd_{0.4}Se_{0.31}Te_{0.19}$ (1) and $Zn_{0.005}Cd_{0.495}Se_{0.31}Te_{0.19}$ (2) samples.

Figure 2 shows PL spectra of vacuum annealed and oxygen doped ZnCdSeTe:Cl:Cu:Pr:Er samples with a different Zn concentration as follows: sample 1 – 10 at%, sample 2 – 5 at%, sample 3 – 2.5 at%, sample 4 – 0.5 at%, sample 5 – without Zn. It seems that a higher Zn concentration allows a higher substitutional solubility of dopants in these alloys and thus, a higher PL intensity. The PL band peak at 1.0 eV is supposedly formed by the A-centre $V_{Zn}Cl_{Te}$, Cl_{Te} and Cu-oxygen complexes [11] whereas a broader 1.2 eV PL band is caused of the host intrinsic defects and bound excitons [12]. Fityk fitting confirmed that the 1.2 eV PL band is formed at least by the 2 Gaussian functions. Oxygen doping increases R_D of the vacuum annealed ZnCdSeTe:Cl:Cu:Pr:Er samples up to 1E11 Ω, corresponding to a resistivity value of about 3E11 Ωcm, and K_F up to 1E6-1E7. It is known that Cu_{Cd} or Cu_{Zn} is a deep acceptor in II-VI compounds whereas Cu-oxygen complexes like Cu_2O and CuO are shallow acceptors that enable a high photoconductivity in these materials or p-type conductivity in CdTe [3, 4]. EDS showed yet Te and $CdCl_2$ traces on the surface of vacuum annealed samples as 1 hr is too short time at 340 ^{0}C to remove all formed precipitates from the thick samples.

Figure 3 depicts PL spectra of two vacuum annealed and oxygen doped $Zn_{0.005}Cd_{0.495}Se_{0.31}Te_{0.19}$ samples having a different Cu concentration: 0.034 at% (sample 1) and 0.05 at% (sample 2). It is seen that a high Cu concentration decreases the PL intensity, especially the 1.2 eV PL band peak. Too high [Cu] also decreases UV sensitivity of the ZnCdSeTe-based radiation detectors, as a high oxygen concentration that forms oxides on the surface of crystals.

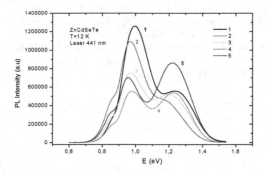

Figure 2. PL spectra of the vacuum annealed 1 hr at 340 ^{0}C and oxygen doped $Zn_{0.1}Cd_{0.4}Se_{0.31}Te_{0.19}$ (1), $Zn_{0.05}Cd_{0.45}Se_{0.31}Te_{0.19}$ (2), $Zn_{0.025}Cd_{0.475}Se_{0.31}Te_{0.19}$ (3), $Zn_{0.005}Cd_{0.495}Se_{0.31}Te_{0.19}$ (4) and $Cd_{0.5}Se_{0.31}Te_{0.19}$ (5) samples.

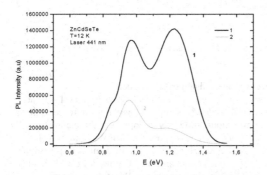

Figure 3. PL spectra of the vacuum annealed 1 hr at 340 ^{0}C and oxygen doped $Zn_{0.005}Cd_{0.495}Se_{0.31}Te_{0.19}$ samples. 1 – with [Cu] = 0.034 at%. 2 – with [Cu] = 0.05 at%.

Figure 4 shows condensed phases formed inside the two-zone vacuum ampoule near the soft shutter of the process tube of furnace during a vacuum annealing 1 hr at 340 ^{0}C of the $Zn_{0.025}Cd_{475}Se_{0.31}Te_{0.19}$:Cl:Cu:Pr:Er samples. Ampoule contained 10 pellets having a total weight about 0.5 g. EDS confirmed that layer No. 1 near 340 ^{0}C is Te which inner surface is in part coated with black TeO. Layer No. 2 is Zn. Dark amorphous layer No. 3 near 300 K contains Te, Zn and Cl, and supposedly consists of Te and Zn chlorides. EDS also detected traces of Cd inside the whole area caused of the condensed $CdCl_2$ crystals that deliquesced in moist air and contaminated this area with Cd. EDS did not detect a separate phase of Se inside the quartz ampoule or on the surface of pellets. Note that a condensed Zn layer No. 2 was formed only in case of the samples which were co-doped with rare earths Pr and Er.

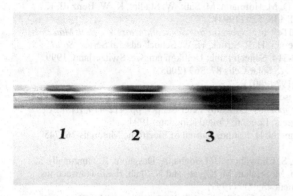

Figure 4. Condensed phases formed inside the two-zone quartz ampoule during a vacuum annealing of the $Zn_{0.025}Cd_{475}Se_{0.31}Te_{0.19}$:Cl:Cu:Pr:Er samples. 1 – tellurium. 2 – zinc. 3 – amorphous mixture of Te and Zn chlorides.

Results of this investigation confirm that Zn in the ZnCdSeTe alloys not only widens the bandgap and increases dark resistivity but also allows a higher substitutional solubility of dopants and as a result, promotes optoelectronic properties of these compounds. Cl dopant substitutes Te in ZnCdSeTe whereas Cu and especially rare earths substitute Zn. ZnCdSeTe semiconductor compound is a good candidate for the resistive radiation detectors from NIR via VIS and UV to X-ray band. These detectors are shown a high stability and good linearity and performance under X-ray excitation of Cu Kα 1.54056 Å, at 40 kV, and withstand electric field up to 2500 V/cm.

CONCLUSIONS

Synthesized chlorine, copper, rare earths and oxygen doped quaternary ZnCdSeTe semiconductor compounds are good candidates for the resistive radiation detectors from NIR via VIS and UV to X-ray band. Zn content in ZnCdSeTe enables to increase dark resistivity of detectors fabricated from this alloy up to 1E11 Ωcm. Zn allows a higher substitutional solubility of dopants in ZnCdSeTe and thus, promotes optoelectronic properties of the ZnCdSeTe alloy. Cl substitutes Te whereas Cu and rare earth elements substitute Zn in ZnCdSeTe.

ACKNOWLEDGMENTS

The financial support by the Bruker AXS, Inc., EU 7th FP project FLEXSOLCELL GA-2008-230861, Estonian Science Foundation grant 7608, and Estonian National Target Financing SF0140099s08 is gratefully acknowledged.

REFERENCES

1. M. Fiederle, D. Ebling, C. Eiche, D. M. Hofmann, M. Salk, W. Stadler, K. W. Benz, B. K. Meyer, Journal of Crystal Growth **138**, 529 (1994).
2. V. Valdna, "p-Type Doping of CdTe", in *Polycrystalline Semiconductors V – Bulk Materials, Thin Films, and Devices*, J. H. Werner, H. P. Strunk, H. W. Schock eds., in Series `Solid State Phenomena` **67-68**, pp. 309-314, Scitech Publ., Uettikon am See, Switzerland, 1999.
3. V. Valdna, Solar Energy Materials & Solar Cells **87**, 369 (2005).
4. V. Valdna, Thin Solid Films **387**,192 (2001).
5. T.-C. Yu and R. F. Brebrick, "Phase diagrams of Cd/Zn/Te/Se compounds", in *Properties of Narrow Gap Cadmium-based Compounds*, Edited by Peter Capper, pp. 412-419, INSPEC, the Institution of Electrical Engineers, London, United Kingdom, 1994.
6. R. Sudharsanan, G. D. Vakerlis, and N. H. Karam, Journal of Electronic Materials **26**, 745 (1997).
7. George C. Giakos, S. Vedantham, S. Chowdhury, J. Odogba, A. Dasgupta, R. Guntupalli, S. Suryanarayanan, V. Vega-Lozada, M. Sridhar, M. Khyati, and N. Shah, IEEE Transactions on Instrumentation and Measurement **48**, 909 (1999).
8. Krihna C. Mandal, Sung Hoon Kang, Michael Choi, Job Bello, Lili Cheng, Hui Zhang, Michael Groza, Utpal N. Roy, Arnold Burger, Gerald E. Jellison, David E. Holcomb, Gomez W. Wright, and Joseph A. Williams, Journal of Electronic Materials **35**, 1251 (2006).
9. S. S. Yoo, G. Jennings, and P. A. Montano, Journal of Electronic Materials **26**, 750 (1997).
10. M. Niraula, K. Yasuda, H. Ohnishi, H. Takahashi, K. Eguchi, K. Noda, and Y. Agata, Journal of Electronic Materials **35**, 1257 (2006).
11. J. W. Allen, Semicond. Sci. Technol. **10**, 1049 (1995).
12. S. Kishida, K. Matsuura, H. Mori, Y. Mizucuchi, and I. Tsurumi, phys. stat. sol. (a) **109**, 617 (1988).

Mater. Res. Soc. Symp. Proc. Vol. 1341 © 2011 Materials Research Society
DOI: 10.1557/opl.2011.1108

Comparative Study of HgI2, PbI2 and TlBr Films Aimed for Ionizing Radiation Detection in Medical Imaging

Marcelo Mulato[1], José F. Condeles[2], Julio C. Ugucioni[1], Ademar M. Caldeira-Filho[1] and Natalia Destefano[1].

[1] Department of Physics, University of São Paulo, Ribeirão Preto-SP Brazil
[2] Department of Physics, Universidade Federal do Triângulo Mineiro, Uberaba-MG, Brazil

ABSTRACT

Wide bandgap semiconductor films were obtained by spray pyrolysis, thermal evaporation and casting. These films were characterized under similar conditions in order to compare their structures, surface morphology and photocurrent properties. All films show either a crystalline or a polycrystalline structure. SEM pictures of sprayed films present holes and fissures and non-total covering of the substrate. The photoresponse was obtained for evaporated TlBr films, HgI2 casted with polystyrene (PS) scaffold, sprayed and evaporated PbI2 films. The photo to dark current ratio is discussed as well as the difference of photo to dark current at an electric field of 100 V/cm. The discussion also focuses on a future optimized material.

INTRODUCTION

Wide bandgap semiconductors, like thallium bromide (TlBr), mercuric iodide (HgI2) and lead iodide (PbI2) have been investigated in last years for X-ray and γ-ray detectors in our research group [1-11]. In search for new techniques to obtain films, we studied techniques like spray pyrolysis (SP), thermal evaporation (TE) and casting (CA). These low-cost techniques could be used to obtain films alternatively to high-cost techniques, which require sophisticated apparatus and are often more complicated to use.

The properties of these materials are high atomic number (Z_{Tl} = 81, Z_{Br} = 35, Z_{Hg} = , Z_I = 53 and Z_{Pb} = 82), high mass density (d_{TlBr} = 7.56 g/cm^3, d_{HgI2} = 6.4 g/cm^3 and d_{PbI2} = 6.2 g/cm^3) and intrinsic band gap (2.68 eV for TlBr, 2.13 eV for HgI2 and 2.4 eV for PbI2). They are interesting for X- and γ-rays detection [1-18].

In this work we fabricated films of these materials and characterized them using structural, morphological and electrical techniques. The main goal of this work is the comparison between these films.

EXPERIMENTAL DETAILS

Films of wide bandgap semiconductors were obtained using SP, TE and CA. The experimental apparatus of these techniques were described in previous works [1-11]. The fabrication conditions are discussed below:
1) TlBr films were obtained by SP and TE. For SP, Mili-Q water was used as solvent and the concentration of TlBr was 1mg/mL. The procedure and parameter of the technique were described previously [1]. Films produced by TE were based on 0.35g of the starting powder that

sits on a tungsten crucible at a home-made TE apparatus [1-3]. We also studied the influence of the number of depositions on the final properties of the films;

2) HgI_2 films were fabricated by SP and CA. Milli-Q water and Ethanol were used as solvent in SP and yellowish and reddish films were obtained respectively [4,5]. The concentration of HgI_2 was 0.06 mg/mL in all studies. Experimental procedures and technique parameters were previously described [4,5]. CA HgI_2 films were obtained with isolated polystyrene (PS) and tetrahydrofuran (THF) as a solvent [6]; and

3) Finally, PbI_2 films were obtained by SP and TE. For SP, dimetylformamide (DMF) and Milli-Q water as solvent following the procedures described in [4,7-10]. TE PbI_2 films were obtained with experimental procedures and technique parameter describe in [11].

The films were characterized by X-ray diffraction (XRD), scanning electron microscopy (SEM) pictures and photo to dark current ratio under X-ray exposure in the mammographic region (30kV and 50mAs - 430mR).

RESULTS AND DISCUSSION

Figure 1 presents the results for the structural analysis. The XRD peaks for all samples were identified by using the Joint Committee on Powder Diffraction Standards. Figure 1a corresponds to a SP TlBr film, which has a polycrystalline structural orientation and the main peak is the (110). However, the TE TlBr film (Figure 1b) has a mono-crystalline structure with the main peak for the (100) direction.

SP HgI2 produced using water (Figure 1c) is yellowish and the main diffraction peak sits at (002) for a orthorrombic net. For production using ethanol instead (Figure 1d) the other diffraction peaks become more important, and a tetragonal structure is obtained.

SP PbI_2 films (either using water or DMF, Figure 1e and 1f, respectively) show similar crystalline orientations with the main peak at (001). However, TE PbI_2 film is mainly oriented along (110) plane. Larger temperature of the substrate heater could favor the appearance of planes at lower 2θ values, as already seem for HgI2 [5]. This happens due to the larger ad-atom mobility on the heated substrate. TE PbI_2 film was performed with substrate at room temperature and the main peak is around 40°.

SEM pictures are presented in Figure 2. Figures 2a and 2b correspond to SP and TE TlBr films respectively. Figure 2c and 2d correspond to SP and CA HgI_2 films respectively, and Figure 2d and 2e to DMF-SP and TE PbI_2 films respectively. Water in spray solution (Figure 2a and 2c) produces films that do not totally cover the substrate for all materials in this study. Although not shown here the same was observed for HgI_2 [5]. Holes and fissures are present, and they could be due to the regions where the excess of solvent accumulates prior to evaporation from the surface of the substrate or to variations of atom mobilities caused by a substrate temperature gradient. A larger deposition time might be used to avoid this problem, but sublimation effects were observed for HgI_2, which prevents the growth of the films [5].

TE TlBr films (Figure 2b) had columnar structure and better covered the substrate than sprayed films. Also the thickness of the films was larger. CA HgI_2 films produces a quasi-smooth surface, showing larger HgI_2 crystals immersed in PS scaffold (Figure 2d). HgI_2 square crystals about 150μm wide were observed. These films had larger thickness than sprayed ones.

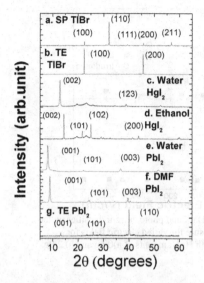

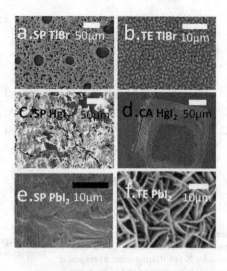

Figure 1 – XRD of a) SP-TlBr; b) TE-TlBr; c) and d) SP-HgI₂; e) and f) SP-PbI₂, and g) TE-PbI₂ films.

Figure 2 – SEM pictures for TlBr [a) SP and b) TE]; HgI2 [c) SP and d) CA]; and PbI2 [e) DMF-SP and f) TE].

SP PbI₂ films with DMF could show porous (showed in work [9]) or denser surfaces (Figure 2e). However, PbI₂ films obtained by thermal evaporation (Figure 2f) have vertical leaves morphology. This structure was previously reported in [11].

Photoelectric measurements were performed for TE TlBr films, CA HgI₂, SP and TE PbI₂ films. Figure 3 presents the results of the photoconductivity under medical X-ray exposure for varied TlBr sequential depositions [2]. The results correspond to the thinner ($n = 1$) and thicker ($n = 4$) samples. Thinner films (Figure 3a) deviated from the linear behavior at approximately 35V (electric field of 5mV/μm). The current saturation above 35 V could indicate that the generated charges are being collected and the increase of applied voltage does not lead to any further improvement. However, thicker samples might have more photogenerated charges, leading to no-saturation current up to 500 V/cm (Figure 3b).

The results of dark and photoresponse under mammographic X-ray illumination for CA HgI₂ films are shown in Figure 4. The films were fabricated using two HgI₂ concentrations of 0.6g (Figure 4a) and 2.0g (Figure 4b). The photoresponse ratio increases with HgI₂ concentration. For increasing PS concentration, the ratio decreases due to a larger separation of crystallites in the PS scaffold (not showed in work [6]).

147

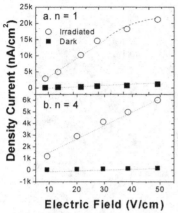

Figure 3 – Dark (squares) and photo(circles) current density of the TlBr films under X-ray illumination: a) response for one single deposition and b) response for four layers deposited.

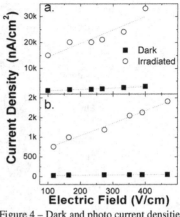

Figure 4 – Dark and photo current densities under X-ray exposure for varied electric fields for samples with HgI_2 amounts of a)0.6g and b) 2.0g.

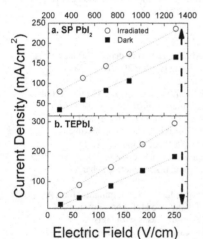

Figure 5 - Dark (squares) and photo (circles) current densities for the films under mammographic X-ray illumination: a) DMF-SP-PbI_2 and b) TE PbI_2.

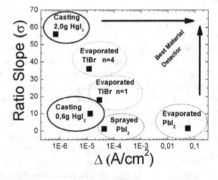

Figure 6 – Ratio of the slopes of the dark and photo currents σ as a function of Δ (difference of current values at 100 V/cm) for all the samples.

Figure 5 presents the current density for SP and TE PbI_2 films under X-rays illumination at the same conditions as TlBr and HgI_2 films. Figure 5a corresponds to SP PbI_2 films which response is linear. Figure 5b) is an electrical measurement for TE PbI_2 film. A linear response is also observed, but with varying parameters.

The slopes of the electrical current density in the dark and under illumination using X-rays were calculated for all samples. It was also calculated the ratio of these slopes (σ) for photo/dark curves. When σ is greater than 1, it indicates that the curves of photocorrent and dark current separate with increasing electric field. When σ is about 1, the curves of photocurrent and dark current are parallel and there is no effective operation window. However, the difference between photocurrent and dark current can be great enough even if the curves ratio is not so high. Therefore, only the analysis of σ values is insufficient to evaluate the material as a photoconductor. Then, we estimated the values of photocurrent and dark current by regression analysis at an electric field of 100V/cm. Their difference (Δ) was calculated. In this sense, we evaluated that the best detector would present both a large value of σ and Δ. Figure 6 shows σ as a function of Δ (A/cm^2) for the samples.

The photoresponse of PbI_2 samples leads to low values of σ and high values of Δ. Therefore, the photo/dark current difference for these films is large compared to TlBr and HgI_2 films, but the samples seem to have a constant σ. The low efficiency of X-ray absorption for these films were related to the thickness of SP PbI_2 [8] and leaves structure for TE PbI_2 [11]. TE PbI_2 presented better values, indicating that this film show a wide range of electric field without photo to dark current variation. TE TlBr films have intermediary values of σ and Δ. The number of deposition (n) influences very little the Δ values, but σ increases twice with the number of depositions. CA HgI_2 films with less HgI_2 amount presents an order of magnitude larger σ values. Of all samples, 2g CA HgI_2 film present the greatest values of σ that indicates this sensor material would be more appropriated to use in high values of electric fields. The best material for use as detector is indicated in Figure 6. We believe that this would be obtained with large energy at the growth surface, combined with larger packing of material when immersed in any scaffold.

CONCLUSIONS

Alternative techniques such as SP, TE and CA were applied for wide bandgap semiconductors to produce films. Different structures were obtained. SP films presented holes and fissures caused by the technique that made it very difficult to obtain dense films. Using other volatile organic solvents instead of water, with a larger degree of dissolution, would be a solution for SP use, as presented for DMF SP PbI_2 film. TE showed different structures for TlBr and PbI_2 films with columnar characteristic. CA HgI_2 films are auto-sustainable and HgI_2 crystals are scattered in PS scaffold. Finally, the photoresponse of films showed diverse results and the films could be applied in different ways. CA HgI_2 films and TE TlBr films showed lower values of Δ, but intermediary and higher values of σ, indicating their use in application with high values of electric field. However, TE PbI_2 films showed higher values of Δ and lower values of σ, indicating the application for a wide range of electric field. We believe that the best material would be obtained with large energy at the growth surface, combined with larger packing of material when immersed in any scaffold.

ACKNOWLEDGMENTS

This work was funded by FAPESP, CAPES and CNPq Brazilian agencies. The authors thank J. L. Aziani, M. Mano Jr, E. A. Navas, S.O.B. da Silva for experimental help.

REFERENCES

1. N. Destefano, M. Mulato. Symp Nucl. Rad. Detec. Mat. 1164, 2009 MRS Spring Meeting, 129 (2010).
2. N. Destefano, M. Mulato. Nucl. Inst.Meth. Phys. Res. A624(1), 114 (2010).
3. N. Destefano, M. Mulato. J. Mater. Sci. 46(7), 2229 (2011).
4. J.F. Condeles, J.C.Ugucioni, M.Mulato. Symp Amorph. Nanocryst. Si Sci. Tech. 808, 2004 MRS Spring Meeting, 489 (2004).
5. J.C.Ugucioni, M.Mulato. J.Appl. Phys. 100(4), 043506 (2006).
6. J.C.Ugucioni, T. Ghilardi Netto, M. Mulato. Nucl. Inst.Meth. Phys. Res. A622 (1), 157 (2010).
7. J.F. Condeles, T.M. Martins, T.C. dos Santos, C.A. Brunello, M. Mulato, J.M. Rosolen. J. Non-Cryst. Sol. 338–340, 81 (2004).
8. J.F. Condeles, T. Ghilardi Netto, M. Mulato. Nucl. Inst.Meth. Phys. Res. A577 (1), 724 (2007).
9. J. F. Condeles, R. A. Ando, M. Mulato. J. Mater. Sci. 43, 525 (2008).
10. J. F. Condeles, M. Mulato. J. Mater. Sci. 46, 1462 (2011).
11. A.M. Caldeira Filho, M. Mulato. Nucl. Inst.Meth. Phys. Res. A636 (1), 82 (2011).
12. T. Onodera, K. Hitomi, T. Shoji, Nuclear Instruments and Methods in Physics Research A 568, 433(2006).
13. P.J. Sellin, Nuclear Instruments and Methods in Physics Research A 563, 1 (2006).
14. J. P. Ponpon, Nucl. Instr. and Meth. A 551, 15 (2005).
15. N. E. Hartsough, S. Iwanczyk, B. E. Patt, and E. L. Skinner, IEEE Trans. Nucl. Sci. 51, 1812 (2004).
16. J. E. Baciak and Z. He, IEEE Trans. Nucl. Sci. 51, 1886 (2004).
17. S.O. Kasap, M.Z. Kabir, J.A. Rowlands, Curr. Appl. Phys. 6, 288 (2006).
18. R.A Street, S.E. Ready, K.V.Schuylenbergh,J. Ho, J.B.Boyce, P. Nylen, K.Shah, L Melekhov, H.Hermon. J Appl Phys 91, 3345 (2002).

Mater. Res. Soc. Symp. Proc. Vol. 1341 © 2011 Materials Research Society
DOI: 10.1557/opl.2011.1109

Photoemission and Cathodoluminescence of Doped Lithium Tetraborate Crystals Being Developed for Neutron Detection

Christina L. Dugan[1], Robert L. Hengehold[1], Stephen R. McHale[1], Yaroslav Losovyj[2,3], John W. McClory[1], and James C. Petrosky[1]

[1]Engineering Physics, Air Force Institute of Technology, Wright Patterson AFB, Ohio
[2]Physics, University of Nebraska, Lincoln, Nebraska
[3]Center for Advanced Microstructures and Devices, Louisiana State University, Baton Rouge, Louisiana

ABSTRACT

Photoemission spectroscopy using synchrotron radiation was used to determine the energy level structure of Mn doped $Li_2B_4O_7$ crystals. Photoemission studies provided evidence of Mn in the bulk crystal at 47.2 eV. Valence band analysis provided the presence of surface states but no acceptor sites. Cathodoluminescence studies were also made on undoped and Mn doped $Li_2B_4O_7$ using various beam energies from 5 to 10 KeV at room temperature. Self trapped exciton emission states are evident in the undoped and Mn doped $Li_2B_4O_7$ sample ranging in energies from 3.1 to 4.1 eV.

INTRODUCTION

In order to better protect the United States against further acts of terrorism within its borders, ports of embarkation, and abroad, smaller, more reliable, sensitive and faster special nuclear material detection capabilities must be explored. Neutron detection has the advantages of low natural background count, few neutron sources in normal commerce operations, and different shielding characteristics compared to gamma rays. Current neutron detection devices are, however, not sensitive enough, bulky, and too expensive for wide use applications.[1] A primary advantage of using solid state media such as semiconductors for detectors is that more dense material yields smaller devices. Solid state neutron device design must utilize a material with a high neutron absorption cross section. Lithium tetraborate, $Li_2B_4O_7$, is a crystalline material containing high densities of ^6Li and ^{10}B, both materials with high neutron absorption cross sections. The ^{10}B neutron capture cross section for thermal neutrons is 3850 barns and ^6Li has a cross section of 940 barns. [2]

The primary objective of this research is to determine the material characteristics of doped $Li_2B_4O_7$ in order to produce more efficient neutron detection devices. Since $Li_2B_4O_7$ is an insulator, it is necessary to dope it with impurities such as manganese, copper, or silver so it may become p type, and thus increase the number of positive charge carriers, or n type to increase the number of electron charge carriers. This would allow for the use of $Li_2B_4O_7$ in heterojunction or homojunction semiconductor devices. An accurate bandmap and electronic configuration of undoped $Li_2B_4O_7$ exists. [3] Through a combination of cathodoluminescence and photoemission spectroscopy the doped $Li_2B_4O_7$ electronic band structure and other surface and bulk characteristics can be determined.

EXPERIMENT

A cathodoluminescence system and two separate photoemission systems were used to study undoped and Mn doped $Li_2B_4O_7$. The cathodoluminescence system consisted of a Kimball Physics EMG-12 electron gun providing above bandgap excitation and a 0.5 m Czerny Turner spectrograph limiting spectral data to a wavelength range of 2000 to 8000 Å corresponding to an energy maximum of 6.4 eV, i.e., approx. 3 eV below bandgap. Photoemission measurements were made on ultraviolet photoemission spectroscopy (UPS) systems at the Louisiana State University (LSU) Center for Advanced Microstructures and Devices (CAMD), using synchrotron radiation on two different beamlines. The first was the 3 m toroidal grating monochromator (TGM) beam line equipped with a 50 mm hemispherical electron energy analyzer providing a resolution of 70 mev and a photon energy range up to 95 eV. The second was the normal incident monochromator (NIM) beamline equipped with a Scienta SES200 electron energy analyzer with a resolution of approximately 10 mev and a photon energy range from 24 to 32 eV to study the band gap states. Measurements were made at elevated sample temperatures to reduce charging effects.

DISCUSSION

Cathodoluminescence Spectroscopy of Li₂B₄O₇

All $Li_2B_4O_7$ crystals have relatively strong broad cathodoluminescence peaks ranging from 3.4 to 3.7 eV. Broad intrinsic emission generally results from self-trapped excitons (STE) or highly localized excitons trapped by their own self-induced lattice distortion. STE's are generally produced in crystals with a deformable lattice characterized by strong electron-phonon coupling. The STE's emission energy is usually much lower than the band gap of the $Li_2B_4O_7$ due to phonon energy lost during the electronic transition.

The cathodoluminescence spectra of the (100) surface of the undoped sample of $Li_2B_4O_7$ was obtained at room temperature with 50 µA of current and a 5 KeV electron beam. This spectrum, shown in Figure 1, has a series of peaks believed to be due to self trapped excitons located at O vacancies or interstitial sites. As the incident beam energy increases, the electrons penetrate further into the sample's surface. The peaks change in relative heights causing a shift in peak strengths. This is caused by a different distribution of STEs; as the energy increased so did the energy of the thermally released STEs. [4]

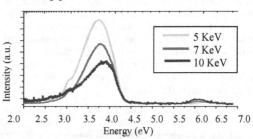

Figure 1. Cathodoluminescence spectrum of (100) $Li_2B_4O_7$ at 5,7, and 10 KeV, 50 uA of current with 400 um slit width, 1 second integration time and a -31°C PMT. 5 KeV energy is the highest intensity spectrum followed by 7, and 10 KeV due to degradation of the sample's surface as time elapsed.

Figure 2 represents the Gaussian fits to the cathodoluminescence spectra of (100) $Li_2B_4O_7$ at 5, 7, and 10 KeV, 50 µA of current, with a 400 um slit width, 1 second integration time and a cooled PMT. It is clear that these spectra are composed of four energy peaks which are at 2.64, 3.05, 3.59, and 3.95 eV. As seen in Figure 1, a weaker intensity peak exists at 6.2 eV probably due to a transition towards the middle of the band gap. This peak is not documented in previous publications. This transition may be between deep donors and the valence band.

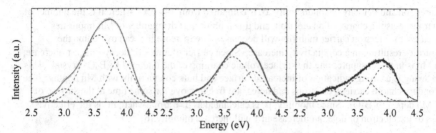

Figure 2. Gaussian distribution of cathodoluminescence spectrum of (100) $Li_2B_4O_7$ at 5, 7 and 10 KeV, 50 uA of current with 1 second integration time, and a -31°C PMT.

The CL spectra of (100) $Li_2B_4O_7$ is in agreement with previous post x-ray irradiation thermoluminescence measurements of undoped $Li_2B_4O_7$ taken at 77-293 K in the AFIT laboratories, shown in Figure 3. The thermoluminescence spectra's energy peaks are at 2.57, 3.10, 3.60, and 4.00 eV in close agreement with those in the CL spectra. At 77K the electrons are trapped at O vacancies and holes are trapped at Li vacancies, both recombining after room temperature annealing. Also, EPR pre and post annealing studies provided additional confirmation that the peaks are attributable to oxygen and lithium vacancies. [5]

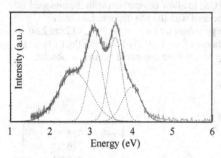

Figure 3. X-ray irradiation thermoluminescence spectra of undoped $Li_2B_4O_7$. Four peak Gaussian fit for the 5 KeV spectra at 2.573, 3.102, 3.579, and 3.993 eV.

Figure 4. Cathodoluminescence spectrum of single crystal Mn doped $Li_2B_4O_7$ at 5 and 7 KeV, 50 uA of current with a -31°C PMT. 5 KeV energy is the highest intensity spectrum followed by 7 KeV due to degradation of the sample's surface as time elapsed.

Doped impurity centers such as Mn provide extrinsic trapping sites as alternatives to self trapping. Mn impurity ions may substitute for lithium ions in the $Li_2B_4O_7$ lattice. The Mn ion impurity likely fills a Li vacancy in the $Li_2B_4O_7$ crystal. The atomic radius of Mn is 1.79 Å and atomic radius of Li is 2.05 Å. The ground state Mn^{2+} ($3d^5$) impurity ions can trap free electrons. The cathodoluminescence spectrum of Mn doped $Li_2B_4O_7$ is shown in Figure 4.

It is possible for the Mn atoms occupying the Li sites to act as donors. Primary collision mechanisms in the lattice are impurity scattering and phonon scattering. Since the spectra are collected at room temperature, both phonon and impurity scattering due to defects will be operative. Collisions resulting from such scattering cause an abrupt change in the carrier velocity and the collision time resulting in carrier motion which is semi-random due to frequent changes

in direction and velocity. Conductivity, σ, is computed from $\sigma = ne
\mu$, where n is the number of electrons, e, is the charge of an electron, and μ is mobility. If the number of Mn impurities increases, the electron carrier mobility will decrease. The increase in n can overcome the impurities resulting in a conductivity increase. Since carrier lifetimes or decay times, τ, increase, the Mn is in effect suppressing the surface voltaic charging of the undoped $Li_2B_4O_7$ crystal. This narrowing is a strong indication of increased surface and bulk conductivity with Mn doping. The secondary charge carriers, electrons and holes, are free to move for some time at the surface of the Mn doped $Li_2B_4O_7$. Since mobility, μ, is calculated as $\mu = e\,\tau/m^*$, where m^* is the effective mass of the electron or hole, an increase in τ will increase mobility.[6]

Note that the Mn doped $Li_2B_4O_7$ CL curve presents a slightly narrower peak at approximately 3.7 eV than the undoped $Li_2B_4O_7$. A narrow peak indicates the exciton carrier lifetime, τ, is much longer for the Mn doped $Li_2B_4O_7$. Narrower cathodoluminescence spectra indicate a strong surface component. The enhanced carrier mobility at the surface is in agreement with greater surface conductivity evident in the reduced surface charging. Hence the increased surface mobilities and carrier lifetimes in the Mn doped $Li_2B_4O_7$ are in agreement with the reduction in hysteresis in the surface voltages which will be observed in the photoemission spectra.

Photoemission Spectroscopy of Mn Doped Li₂B₄O₇

Photoemission spectra were taken with the TGM to allow comparison with the undoped $Li_2B_4O_7$ UPS spectra and to determine energies associated with the Mn dopant. Literature reviews and previous publications have binding energy values for Li $1s$ at 54.7 eV, O $2s$ at 22.0 eV, and Mn $3p_{1/2}$ and $3p_{3/2}$ at 47.2 eV with an uncertainty of +/- 1 eV.[7] To ensure the Li $1s$ and O $2s$ core energy levels were correct, the core energy levels were compared against an absolute theoretical value using least squares fitting.

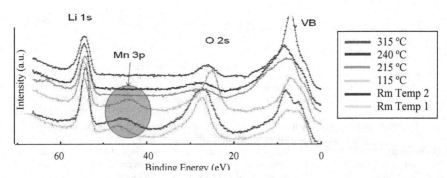

Figure 5. Binding energy versus intensity for Mn doped Li₂B₄O₇ with 110 eV photons at various temperatures. Spectra are arranged from highest to lowest temperature.

Figure 5 shows the UPS curves taken at different sample temperatures. One sees that as the temperature increases, the photoemission peaks shift to the left. In the case of the valence band, the thermal energy helps mitigate the valence band shift by increasing the number of charge carriers and their mobility diminishing the positive surface charge. That is, the increase in temperature created thermal energy for electrons from deeper in the sample to move towards the empty orbitals near the surface of the sample. This mitigates the surface charging effect and provides spectra that are closer to ground state reality.

Again referring to Figure 5, one sees that as the temperature increases from room temperature to 215 °C the Mn peak disappears, whereas, when the sample was cooled from 240 to 115 °C the Mn peak reappeared. The Mn peak was only present below 115 °C. Also as the sample is heated, the surface charges and the valence band changes. The O $2s$ levels are affected, but the deep Li $1s$ level remains quite constant since Li is an alkali metal, which, when singly ionized, possesses a noble gas configuration. The second ionization potential of lithium is thus quite high, due to the difficulty in removing any electron from an atom having a noble gas configuration. Hence, lithium atom 1s electrons are not easily ionized.

Photoemission spectra also were taken using the NIM beamline with a 0.01 degree resolution Scienta energy analyzer providing an energy resolution of approximately 10 meV for photons with incident energies varying from 24 to 32 eV. The purpose of the NIM measurements was to examine the region of the valence band edge to the center of the bandgap for acceptor states. At these energies, the photons can only excite transitions from the acceptor states of Mn doped $Li_2B_4O_7$ near the valence band.

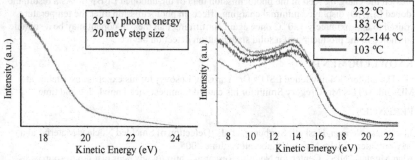

Figure 6 a.NIM UPS spectra with 26 eV photon energy at 20 meV step size. b.Temperature dependence of 26 eV photon energy. Spectra are arranged from highest to lowest temperature.

Figure 6a is the resulting intensity versus kinetic energy graph. As was stated earlier, this long, detailed scan was conducted to see if any acceptor states were present. As can be seen from the data, no clearly discrete acceptor states were found; the only possible acceptor states would be located above 17.75 eV, a region of the PES curve which is very smooth. As was stated earlier, the smooth structure in this region is probably due to surface states. Finally, a temperature dependent scan was conducted at 26 eV photon energy to determine what sample temperature is required to adequately reduce the surface charging of the Mn doped $Li_2B_4O_7$. The second graph in Figure 6b represents the intensity versus kinetic energy curve of Mn doped $Li_2B_4O_7$ at various temperatures. As can be seen from this second graph, charging was significantly reduced at temperatures greater than 122 °C. This reduction in charging is believed to be due to the presence of Mn in the crystal.

The photoemission spectra are in agreement with the cathodoluminescence spectra demonstrating a reduction in surface voltaic charge. As the photoemission spectra increase in temperature and then return to room temperature, the surface voltaic charging indicates only a little hysteresis even though this material is pyroelectric. In comparison with photoemission spectra from the undoped and Cu doped $Li_2B_4O_7$, the Mn doping appears to have suppressed the charging and the pyroelectricity. [8] The Mn doped $Li_2B_4O_7$ has a (001) crystal orientation, i.e. the pyroelectric direction, and seems to produce less surface charging than the undoped and Cu doped $Li_2B_4O_7$; it may be the most promising dopant for a $Li_2B_4O_7$ solid state neutron detector.

CONCLUSIONS

The cathodoluminescence spectra provided evidence of self trapped excitons in both the undoped and Mn doped $Li_2B_4O_7$. As well, the narrowing of the primary cathodoluminescence peak around 3.7 eV in the Mn doped $Li_2B_4O_7$ confirmed Mn doping increases the conductivity of $Li_2B_4O_7$ enhancing its potential as a solid state neutron detector. The TGM photoemission spectroscopy was paramount to this study as it was these measurements that confirmed the presence of Mn in the samples. Interestingly, the Mn ions were found to segregate to the surface at low temperatures and dissolve into the bulk of the material at the higher temperatures. The NIM photoemission spectroscopy established the presence of surface states but revealed no acceptor states. Also, there is strong indication that the presence of Mn in the Mn doped sample suppressed the charging seen in the photoemission data of the undoped $Li_2B_4O_7$. As a result, the Mn doped $Li_2B_4O_7$ displayed minimal charging effects in photoemission after the temperature was raised to approximately 130 ^{0}C once again confirming that the Mn dopant may be a suitable initial choice in the production of a solid state neutron detection device.

ACKNOWLEDGMENTS

The authors acknowledge LSU's Dr. Yaroslav Losovyj for his expertise assistance at CAMD and AFIT's Mr. Gregory Smith for his cathodoluminescence knowledge and time.

REFERENCES

[1] J.L. Jones, Norman, Haskell, Sterbentz, et al. "Detection of Shielded Nuclear Material in a Cargo Container", Idaho National Laboratory, June 2005.
[2] Alan Munter. "NIST Center for Neutron Research," http://www.ncnr.nist.gov/resources/n-lengths/, (2003).
[3] B. V. Padlyak, Y. V. Burak, and V. M. Shevel, "Neutron -Induced Defects in the Lithium Tetraborage Single Crystals," Radiation Effcts & Defects in Solids, 1101-1109, (2002).
[4] Volodymyr T. Adamiv et al., "The Electronic Structure and Secondary Pyroelectric Properties of Lithium Tetraborate," Materials, vol. 3, 4550-4579,(2010).
[5] M. Swinney, Defect Charactersization, "Scintillation Properties, and Neutron Detection Feasiblity of Lithium Tetraborate", 2010.
[6] Adamiv, V. T., Wooten, D. J., McClory, J., Petrosky, J., Ketsman, I., Xiao, J., et al., "The Electronic Structure and Secondary Pyroelectric Properties of Lithium Tetraborate," Materials, 3, 4550-4579 (2010).
[7] Stefan Hufner, *Photoelectron Spectroscopy Principles and Applications*. New York: Springer, 2003.
[8] David J. Wooten, "Electronic Structure of Lithium Tetraborate," Air Force Institute of Technology, Wright Patterson AFB, 2010.

AUTHOR INDEX

SUBJECT INDEX

Bi, 125

chemical synthesis, 87
crystal, 9, 21
crystal growth, 21, 29, 39, 45, 95,
    111, 139
crystallographic structure, 145

defects, 9, 61, 67
densification, 3
devices, 61
dislocations, 61
dopant, 9
Dy, 105

electrical properties, 145
electronic material, 139
electronic structure, 81

film, 145

Gd, 75

II-VI, 29, 39, 45, 67, 139
In, 95

laser, 15
laser-induced reaction, 15
liquid-phase epitaxy (LPE), 119

luminescence, 15, 119, 151

nanoscale, 51, 105
neutron irradiation, 75, 105, 151
nuclear materials, 21, 87, 133

optical properties, 133
oxide, 119

Pd, 51
photoconductivity, 133
photoemission, 151
polymer, 125
porosity, 3
purification, 45

radiation effects, 87

semiconducting, 29, 39, 95
sensor, 111, 125
simulation, 75, 81
sintering, 3

transmission electron microscopy
    (TEM), 51, 67
transportation, 81

zone melting, 111

Printed in the United States
by Baker & Taylor Publisher Services